Intermolecular Interactions

Intermolecular Interactions

Edited by

Werner Gans

Free University of Berlin
Berlin, Germany

and

Jan C. A. Boeyens

University of the Witwatersrand
Johannesburg, South Africa

SPRINGER SCIENCE+BUSINESS MEDIA, LLC

Library of Congress Cataloging in Publication Data

Intermolecular interactions / edited by Werner Gans and Jan C. A. Boeyens.
    p.   cm.
"Proceedings of the Second Structural Chemistry Indaba on Intermolecular Interactions, held August 3–8, 1997, in Kruger [National] Park, South Africa"—T.p. verso.
Includes bibliographical references and index.
ISBN 978-1-4613-7189-2   ISBN 978-1-4615-4829-4 (eBook)
DOI 10.1007/978-1-4615-4829-4
    1. Molecular dynamics—Congresses. 2. Chemical structure—Congresses. 3. Intermolecular forces—Congresses. I. Gans, W., 1949–   . II. Boeyens, J. C. A. (Jan C. A.) III. Structural Chemistry Indaba on Intermolecular Interactions (2nd: 1997: Kruger National Park, South Africa)
QD461.I574   1998
541.2'26–dc21   98-28412
                                                                                                        CIP

Proceedings of the Second Structural Chemistry Indaba on Intermolecular Interactions, held August 3 – 8, 1997, in Kruger Park, South Africa

ISBN 978-1-4613-7189-2

© 1998 Springer Science+Business Media New York
Originally published by Plenum Press,New York in 1998
Softcover reprint of the hardcover 1st edition 1998

http://www.plenum.com

10 9 8 7 6 5 4 3 2 1

PREFACE

This volume contains most of the invited lectures of the 2nd Structural Chemistry Indaba on "Molecular Interactions," held at Skukuza, Kruger Park, South Africa, August 3–8, 1997. While the 1995 conference concentrated more on the principles underlying molecular modeling, like the existence of a molecular shape, this conference centers on molecular interactions or, more generally, on molecules in environments.

Unfortunately, it was impossible, for various reasons, to unite all invited lectures in this volume, but nevertheless this collection contains contributions ranging from the fundamental quantum mechanical theory to recent research on organometallic crystals. For a summary, I would like to refer the reader to the introductory chapter by S.O. Sommerer, based on his concluding remarks at the conference.

Werner Gans
for the editors

CONTENTS

Intermolecular Interactions

INTERMOLECULAR INTERACTIONS

Shaun O. Sommerer

Department of Physical Sciences
Barry University
Miami Shores, Florida 33161

Intermolecular interactions is a difficult topic to probe not because it is a rare phenomenon or so subtle that it escapes detection but rather because of varied opinion among scientists as to what exactly constitutes "a molecule." Most structural chemists who investigate the architecture of matter at the picometer level ordinarily presuppose that the concept of "a molecule" is well established in a fashion similar to that of the periodic chart and if asked, happily sketch out several examples that they are currently characterizing. A summary of such responses provides a rather sweeping range of examples (i.e. definitions); moreover, if responses from material scientists, physicists or molecular biologists are included, an impressive spectrum is revealed limited by the few who argue at the extremes that no definition of "a molecule" adequately communicates anything useful about the structure of matter. Diversity such as this provides a compelling argument for a concerted effort towards establishing a consensus as to what "a molecule" is before meaningful discussion can begin regarding interactions between molecules. Indeed, when a concept as central as that of "a molecule" becomes increasingly difficult to pin down, it becomes imperative for a group of scientists with diverse backgrounds to gather and analyze the problem from a variety of angles. In Africa such a gathering is called an Indaba.

Just exactly what a molecule is or is not depends upon the question one is setting out to answer. The definition of "a molecule" one would find useful if investigating conglomerate crystallization is quite different than that employed by a computational chemist interested in kinetics of catalysis. More importantly however are the presuppositions one unavoidably brings with when they set forth a hypothesis and design an experiment. Without question volumes of excellent science has been reported but for practical purposes be they cost, time or instrumental limitations, every report carries with it preconceptions deeply imbedded within the scientific enterprise. Such concepts are extremely useful and necessary as long as an experiment remains within a particular horizon of understanding but misunderstandings abound when science done under one rubric is integrated into another horizon or is made the interlocking link to an anticipated wider horizon. The utility of concepts previously thought to be well established facts lose clarity to the point, in some cases, of actually impeding progress towards new knowledge. For example, an organometallic chemist who prepares a metal-containing-polymer may cling desperately to the notion of the formula unit as being "a molecule" and the extended chain as being composed of many of these molecules where the polymer chemist sees the chain as the molecule. The propensity for language difficulties mounts when the discussion of intermolecular interactions ensues; however, insofar as we have some experience and understanding of "a molecule" no matter how abstract, phenomena described as intermolecular interactions can be appreciated.

The major research horizons converging within this collection of papers providing glimpses of such interactions can be broadly grouped into three areas: experimental, theory/computational and data mining. Each of these groups interfaces with the other two as evidenced by the many current papers reporting or making use of results acquired from the other two horizons. There is however, a fourth horizon that is essential to any fundamental, in-depth examination and that is the philosophic horizon in the mode of scientists observing scientists doing science. If the motivation for an Indaba is to examine fully a fundamental concept, then it is of paramount importance to include voices that examine, critique and challenge the origins of concepts assumed to be fact. Taken together, the four horizons provide a three-dimensional framework for an Indaba just as the four vertices of a tetrahedron make it a three-dimensional object. It is tempting to think that little more than confusion could result from so diverse a discussion: that one would come away with few insights and be left with nothing other than a bad taste and a renewed confidence in primitive models. Quite the opposite is the case. Many rewards await those willing to enlist in a discussion of fundamentals but effort is required together with a good portion of humility.

Although it is timely to undertake and examination of intermolecular interactions it is important to realize that no matter which research horizon we are part of, there comes a point where we have to make some choices so as to bring the project to some closure. Recognizing that it is only in exhaustive questioning of accepted presuppositions that progress can be made and new insights developed, it is constructive to discuss and present evidence of intermolecular interactions from each of the horizons in order to broaden and expand one's personal horizon which ultimately will lead to deeper insights. This is correctly understood as a process and one quickly recognizes the interdependency of such an undertaking. The data minors make all contributors to the process in some degree so it is imperative that investigations continue to be carried out in responsible fashion meaning that we are attentive as we collect data, that we are reasonable in determining our limitations of what can and cannot be done. If we operate authentically at these levels in our basic investigations we can then make true progress even if it is only the realization that our current definitions are exhausted. Most likely one will come away with a more integrated vision of intermolecular interactions which sees this phenomena as a high point in the continuous drama staged by the basic constituents of matter.

INTERMOLECULAR BONDING

Jan C.A. Boeyens

Centre for Molecular Design
Department of Chemistry
University of the Witwatersrand
Johannesburg

1. INTRODUCTION

Of all the elusive concepts in chemistry, the very fashionable notion of an "intermolecular interaction" could well be one of the more ambiguous. It presupposes a clear understanding of an isolated molecule in relation to its environment, commonly made up of similar molecules. To achieve this it is necessary to ignore the very interactions that one aims to describe. This dilemma stems from the fact that molecular shape is a function of the environment. On the one hand the molecule therefore acquires shape through interaction with the environment, and on the other, interactions with the environment are inferred from an assumed intrinsic molecular shape.

An extreme example is provided by the solution of an amphiphilic species which undergoes dramatic changes at a critical concentration, with spontaneous assembly into globular micelles. These globules change their shape when the polarity of the medium is adjusted, and could eventually invert themselves[1]. At intermediate polarities a solution of amphiphilic monomers is formed. Crystallization could very likely produce different crystalline modifications, depending on the polarity. In each of the more than five states of aggregation the notion of intermolecular interaction will have a different meaning and the definition of what constitutes the "molecule" in each case will most likely be in dispute.

In solutions above the critical micelle concentration it may seem logical to consider the globular aggregates as macromolecules, but then it could be argued that the monomeric units are not held together by covalent bonds[2]. The problem with this is that covalent bonds are not in a simple class by themselves. An instructive exception is provided by ferrocene, which is not stabilized by bonds in the Lewis sense either. Although the binding is probably more akin to micellar cohesion than to common electron-pair bonding, the molecularity of ferrocene is never in doubt.

It would be wrong to infer that the distinction between inter- and intramolecular interactions becomes blurred only in some esoteric special cases. A recent survey[3] has in fact identified partially bonded molecules from the solid state to the stratosphere. Conclusions were largely drawn from a comparison between gaseous and crystal-phase structures. One of the many examples is the complex $HCN \cdot BF_3$, which is a solid with high vapour pressure at room temperature. Comparison by microwave spectroscopy and X-ray diffraction showed that on crystallization the N–B bond length changes from 2.47 Å to 1.64 Å, while the B–N–F angle increases from 91.5° to 105.6°. The general conclusion was that the transition between Van der Waals and chemical interactions was continuous and only dependent on the environment. In the case of the N–B bond this covers a range from 2.88 Å, in the Van der Waals complex, to the 1.58 Å covalent bond. This is an example of an intermolecular interaction smoothly converting into an intramolecular interaction.

A more familiar example of how molecules respond to the environment is found in the so-called structure-correlation studies [4], originally designed to correlate crystallographic intermolecular contacts with chemical reaction pathways. In fact, it illustrates how the same interatomic contact can be viewed as changing continuously, from describing an intermolecular into describing an intramolecular interaction. Compelling experimental evidence therefore suggests that chemical bonds are not immutable, that the definition of a molecule in terms of chemical connectivity is not unique, and that the distinction between intra- and intermolecular interactions is an arbitrary one.

2. THEORETICAL CONSIDERATIONS

The ambiguity between inter- and intramolecular interactions exists because of an artificial distinction, usually drawn between different types of chemical bond. This is contrary to experimental evidence that reveals a continuous range of interactions, spanning the various kinds of bond. It is therefore of interest to examine the theories of chemical bonding with an eye on the common features of interactions, traditionally considered to be of different origin. The real issue is to identify the fundamental differences, if any, between covalent, ionic, dipolar, dative, dispersive, hydrophobic, metallic, and any other type of chemical interaction. The first question to settle is how a chemical interaction, of whatever kind, comes about in the first place, starting from atoms as the ultimate elementary building blocks.

The magnitude of an interatomic interaction depends on two factors: the difference in chemical potential, historically known as chemical affinity, and the influence of the environment. The chemical potential depends on the electronic structure of the atom only, and for this reason it also changes in response to environmental factors. Comparison of different atomic affinities therefore requires the definition of a standard state for chemical interactions, but this has never been attempted. The so-called *valence state* is used instead, and this is characteristic of each atom. Valence-state affinity, also called electronegativity[5], as an index for chemical reactivity therefore has the disadvantage of referring to different conditions for different atoms.

To understand how an atom gets promoted into the valence state it is only necessary to consider the response of its electronic charge cloud to isotropic compression. This is readily simulated by calculating the quantum-mechanical radial electronic probability density, using an adjustable boundary condition on the the wave function[6]

$$\lim_{r \to r_0} \psi = 0 \tag{1}$$

for $r_0 < \infty$. The effect of this compression is to raise all electronic energy levels of the atom. As the excess-over-ground-state energy is transferred between electrons, the valence state is reached when one electron gains enough energy to decouple from the atomic core. This clearly happens at a different characteristic value of r_0, known as the ionization radius, for each atom. It means that there is no common or standard valence state. Different atoms enter the valence state under different conditions of pressure, temperature and concentration. Because of their different electronic configurations even atoms with the same ionization radii do not necessarily require the same energy of compression to reach their valence states.

An atom in its valence state consists essentially of a free electron confined within a sphere, defined by the ionization radius (R), and decoupled from the monopositive core. The energy of this valence electron is given by

$$E = \frac{h^2}{8mR^2} \tag{2}$$

and, being at the highest atomic energy level it therefore represents the Fermi energy, or chemical potential for the atom in the valence state. This is commonly referred to as the electronegativity[7] of the atom. Whereas these electronegativities are clearly not reduced to a common standard state, it is unreasonable to expect that they could quantify the course of a chemical reaction involving different kinds of atom. As known from experience it provides, at best, only a qualitative measure of chemical affinity.

3. CHEMICAL INTERACTION

The simplest type of chemical interaction takes place in a single-phase medium when promoted into the valence state by external factors. In this case all reactants (atoms) reach the valence state under the same conditions. In this excited or reactive state each reactive unit consists of an electron decoupled from its parent core and the interaction that follows takes place between these itinerant electrons, set free to spread across the reaction system[9].

The most likely interaction, in the first instance, is with neighbouring reactive units to form electron-pair bonds (diatomic molecules), which may, or may not, react further to form oligomers (like S_8 or C_{60}), depending on the valence state of the binary units. Complete reaction results in the formation of a crystalline supermolecule,[2] like diamond, or a metal. Intermediate stages produce identifiable products with lower symmetry and in which more than one type of neighbouring contact (interatomic distance) occur. These are the contacts said to characterize intra- and intermolecular interactions.

From another point of view the secondary contacts are precisely what were identified before[3] as partial bonds in species like $HCN \cdot BF_3$. In this sense intermolecular interactions are no more than manifestations of incomplete chemical reaction. Where the valence states of the intermediates lie close to that of the single unit, complete reaction occurs. If not, the intermediates are identified as molecules, provided the primary interaction is irreversible under ambient conditions. It is noted that the reversibility depends on the activation barrier which characterizes the course of the reaction, as shown in figure 1[8]. The activation barrier is nothing but a measure of the valence state for the reaction.

For interactions between dissimilar atoms the situation is more complicated, since the interacting units now have different valence states. First reaction occurs when one of the atoms, usually the more electropositive, reaches the valence state and releases an electron. If there is a large discrepancy in promotion states, this valence electron is effectively transferred to an atom of the second kind and ionic interaction occurs.

Formation of a covalent bond now happens only under conditions, sufficiently severe for both species to reach their promotion states and on return to milder conditions this interaction would most likely reverse itself. At all intermediate situations partial transfer of charge between the two entities at different chemical potentials takes place. This corresponds to the Sanderson[10] principle of electronegativity equalization. The resulting bonds are all of the same partial type mentioned before.[3] Since the difference in chemical potential is a function of the environment the bonding geometry will likewise be sensitive to the state of aggregation and other thermodynamic factors.

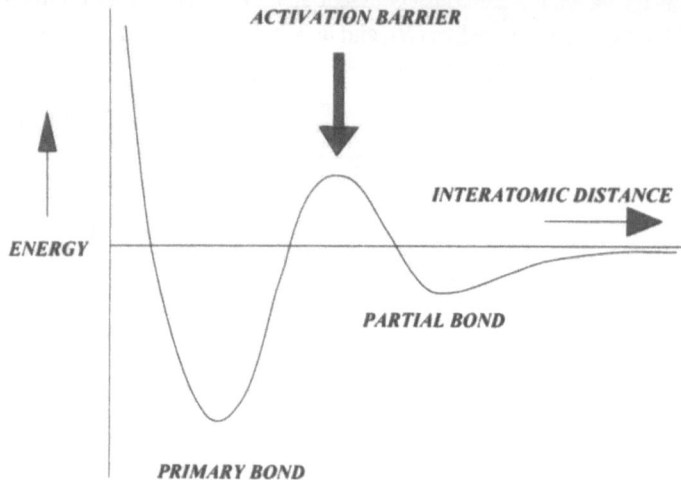

Figure 1. Schematic diagram showing the course of an interatomic interaction as a function of interatomic distance. The height of the activation barrier determines the reversibility of the bond formation.

4. THE MOLECULAR STRUCTURE HYPOTHESIS

A molecule of fixed geometry implies a fixed connectivity pattern. As illustrated in the foregoing, precious few arrangements can meet this requirement, unless the thermodynamic conditions are well defined. Since each compound has its own phase relationships and region of stability this thermodynamic molecular state will be unique to each compound. A common molecular state can therefore not be defined more generally than in terms of an arbitrary invariant point in phase space. Likely consensus on where to choose this point is best left to the imagination of the reader.

Many chemical compounds in the crystalline state may however, be characterized in terms of a three-dimensional robust array of atomic nuclei in a sea of electron density. Concepts like bond length, valence angle, torsion angle, molecular structure, conformation, molecular planarity, intermolecular contacts and a host of other compound-specific concepts like steric congestion, cone angle, coordination geometry, and polyhedron sharing are parameters derived by the chemist from the nuclear arrangement in order to highlight chosen aspects of the cohesion pattern. In the more fluxional liquid, solution or gaseous states, these concepts largely loose their meaning, but to facilitate speculation about changes of state, they are often retained to rationalize trends in observed physical properties like viscosity.[11] This is the essence of the molecular structure hypothesis.

An alternative is to discard the fixed-structure hypothesis and to accept that the molecular identity of any compound depends on its thermodynamic state. Rather than specify a molecular structure for a given compound one would then detail a geometrical arrangement for each state of interest. This description would typically include specification of a more-or-less covalent core and the partial bonds linking these cores into a larger structure. The important feature of this approach is that both core and partial-bond connectivities may differ between states. Since different interactions come into play at each phase transition,[11] the identity of logical units defining notional molecules in different phases, must be different.

It becomes immaterial whether the phases are distinguished in terms of different inter- or intramolecular interactions. The only real difference is that the latter choice is not allowed under the molecular-structure hypothesis, which postulates an invariant core, not affected by changes of state. This proposition is hard to defend under all circumstances and maybe one should resist the grafting of recognizable features from the crystalline state into other environments. This leaves minimal room for discourse with respect to the presumed structure of fluxional or disordered states.

5. STRUCTURE REARRANGEMENT

As a working hypothesis it could be assumed that nuclear motion is restricted, although not eliminated, in the solid state. The crystallographically observed static arrangement, in this sense, represents a minimum in the chemical potential of any compound, which is a function of temperature and pressure. A given solid phase represents the situation that best smooths out the the differences in atomic chemical potential (electronegativity) for that state. In a diamond the potential field should, for instance, be almost featureless.

Changes in temperature and pressure produce different responses in the atomic chemical potentials of different types of atom and hence may cause rearrangement of the nuclear framework. Each state requires a different configuration to minimize the total chemical potential. This scenario ignores metastable situations resulting from activation barriers which render certain rearrangements irreversible.

Different solid phases appear because of changes in kinetic energy. As a function of increasing temperature (and/or reduced pressure) rotation of molecular fragments sets in, followed by the appearance of localized vibrational modes and translational motion at the melting point. Interactions existing in the most ordered solid phase are overcome by kinetic energy in the sequence of their increasing bond strengths or diminishing activation barriers. Elimination of each type of interaction produces a new phase of higher symmetry and reduced structure. There is no common molecular shape that occurs across all phases and no unique set of intramolecular interactions either.

One is left with a simple picture, not too far removed from the traditional distinction between intra- and intermolecular interactions, but based on different premises. The distinction is now made on the basis of reversibility of bonding. Strong bonds, not easily ruptured and therefore irreversible in the present context, usually survive phase transformations and include most of the traditional intramolecular interactions. More reversible, weaker bonds broadly replace the traditional intermolecular interactions and partial bonds.[3] The important new features of this scheme is that a continuous range of bonds, from irreversible to fully reversible, is recognized and that this hierarchy is not fixed, but variable with changes of state.

REFERENCES

1. D. Attwood and A.T. Florence, "Surfactant Systems", Chapman and Hall, New York (1983).
2. F.H. Allen, This volume.
3. K.R. Leopold, M. Canagaratna and J.A. Phillips, *Acc. Chem. Res.* 30:57 (1997).
4. H.B. Bürgi and J.D. Dunitz, eds., "Structure Correlation," VCH Publishers, Weinheim (1994).
5. J.C.A. Boeyens and J. du Toit, unpublished (1997).
6. J.C.A. Boeyens, *J. Chem. Soc., Faraday Trans.* 90:3377 (1994).
7. R.G. Parr and R.G. Pearson, *J. Am. Chem. Soc.* , 105:7512 (1983).
8. S.D. Travlos and J.C.A. Boeyens, *S. Afr. J. Chem.* 50:17 (1997).
9. J.C.A. Boeyens, *Electr. J. Theor. Chem.* 1:38 (1995).
10. R.T. Sanderson, "Chemical Bonds and Bond Energy," Academic Press, New York (1971).
11. J.C.A. Boeyens, in W. Gans et al., eds. "Fundamental Principles of Molecular Modeling", Plenum Press, New York, p. 35 (1996).

CHEMICAL REACTIONS IN THE FRAMEWORK OF SINGLE QUANTUM SYSTEMS

Anton Amann

Universitätsklinik für Anästhesie und Allgemeine Intensivmedizin
Leopold-Franzens-Universität Innsbruck
Anichstr. 35
A–6020 Innsbruck

ABSTRACT

Quantum mechanics in its traditional form is adapted to describe ensembles of identical systems (with a density-operator formalism including dissipation) or single isolated systems. But the traditional formalism is *not* capable of describing single *open* quantum objects with many degrees of freedom showing pure-state stochastic dynamical behaviour. Similarly, *at best* chemical reactions between *ensembles* of molecules can be described in a rigorous manner (including energy dissipation), because dissipation of energy between a single quantum object (e. g., consisting of molecules) and an environment (e. g., the radiation field) always runs via non-product states which entangle object and environment. Usually, this problem is overcome by introducing "quantum jumps" à la Bohr: typically, a molecule and the radiation field jump simultaneously between energy eigenstates, e. g., the molecule emitting energy $E = E_2 - E_2$ and the field mode with frequency $\nu = E/h$ absorbing E by changing its energy eigenstate (number state of the respective quantum harmonic oscillator) accordingly. Or, in the context of chemical kinetics, a particular family Φ_ν, $\nu = 1, 2, \ldots$ is declared to comprise all possible final states of some chemical reaction, an assumption which then allows to introduce and compute reaction rates k_ν.

From a fundamental point of view, this is not satisfactory, because a stochastic dynamics with "quantum jumps" or "instantaneous transitions" between different states cannot rigorously be derived from a Schrödinger-type (continuous) automorphic dynamics. Though (Fermi Golden Rule) transition probabilities are derived by handwaving arguments (and detailed, perhaps fastidious calculations) using an automorphic Schrödinger dynamics in quantum field theory, no rigorous understanding for such chemical reactions has been gained.

Here we try to encorporate (continuous) pure-state stochastic dynamics aimed for a rigorous description of single quantum systems (instead of ensembles of quantum systems, as, e. g., ensembles of molecules). Our generic example of a stochastic dynamics is stochastic "line-jumping", experimentally observed in single-molecule spectra of defect molecules in a molecular matrix. There is no experimental indication that this "line-jumping" is discontinuous. Nevertheless we start with a discussion of the Bohr scenario, which uses stochastic quantum jumps (between strict molecular energy eigenstates) to interpret single-molecule spectra. We use the Bohr scenario as our starting point to try to encorporate more general pure-state stochastic dynamical behaviours into the quantum mechanical formalism.

Such more general pure-state stochastic dynamics should give insight into the detailed processes explaining chemical reactions. In particular, one should try to understand how reactant molecules get quantum-mechanically entangled and react with a successive disentanglement of the reaction products and explain how particular families of final states arise. Also, stochastic external control should allow to choose specific reaction channels and hence guide a chemical reaction into one particular final state.

1. MODELLING CHEMICAL REACTIONS IN QUANTUM MECHANICS

We start with a chemical reaction of a few reaction partners, such as

$$A + B = C + D \tag{1}$$

or

$$A = B \tag{2}$$

From a chemical point of view, such reactions are simple, but not so in quantum mechanics. Here "simple" does not refer to thermodynamic or other calculations, but to the conceptual background which is entirely different in chemistry and quantum theory. To illustrate this statement, we shall try to give some description in terms of the Bohr scenario, i. e., by using (approximate) energy eigenstates of the reactants and products

$$\hat{H}_A \, \psi_{A,j} = E_{A,j} \, \psi_{A,j} \tag{3}$$
$$\hat{H}_B \, \psi_{B,j} = E_{B,j} \, \psi_{B,j} \tag{4}$$

$$\cdots$$

These (approximate) energy eigenstates are (approximately) stationary with respect to the time-dependent Schrödinger equation, i. e., the respective wave vectors $\psi_{.,j}$ do not change in time. If we bring two molecules (of species) A and B together, then things will change: with respect to the joint system consisting of molecules A and B with hamiltonian $\hat{H}_{A,B}$ the product state

$$\psi_{A,j} \otimes \psi_{B,j} \tag{5}$$

is *not* stationary (for the moment, we ignore additional problems like the Pauli principle). Hence the Schrödinger time-evolution will lead to nontrivial, entangled states

$$(\psi_{A,j} \otimes \psi_{B,k})_t = \exp\left\{-i\hat{H}_{A,B}t/\hbar\right\} (\psi_{A,j} \otimes \psi_{B,k}), \tag{6}$$

where "entangled" means that the time-evolved state in eq. (6) cannot be viewed any more as a product state.

Chemical experience tells us that (after some reaction time t_{reaction}) we get new species C and D. Translated into the language of quantum mechanics, this means that we have approximate product states

$$\exp\left\{-i\hat{H}_{A,B}t_{\text{reaction}}/\hbar\right\} (\psi_{A,j} \otimes \psi_{B,k}) = \psi_C \bar{\otimes} \psi_D. \tag{7}$$

Using the Bohr picture again, we would like to end with (approximate) eigenstates $\psi_{C,m}$ and $\psi_{D,n}$ of the new species C and D, i. e.,

$$\exp\left\{-i\hat{H}_{A,B}t_{\text{reaction}}/\hbar\right\} (\psi_{A,j} \otimes \psi_{B,k}) = \psi_C \bar{\otimes} \psi_D \overset{?}{\rightarrow} \psi_{C,m} \bar{\otimes} \psi_{D,n}. \tag{8}$$

The hamiltonian $H_{A,B}$ coincides with the hamiltonian $H_{C,D}$, since it only depends on the over-all number of electrons and nuclei, which is identical for reactant and product species. Note that two different tensor products denoted by $\otimes$ and $\bar{\otimes}$ are used here, which tensor the joint system of electrons and nuclei into reactant species A and B or product species C and D, respectively.

Translating chemical reactions into quantum mechanics in this way, several problems arise: Starting with product states $\psi_{A,j} \otimes \psi_{B,j}$, how does one get to product states $\psi_C \bar{\otimes} \psi_D$ with respect to the reaction product species? And secondly, keeping to the Bohr scenario, how do we get to (approximate) eigenstates $\psi_{C,m} \otimes \psi_{D,n}$ for the product species C and D?

One could, and does often, treat the nuclei as behaving *classically*: this means that one has always product states with respect to (all) the nuclei, and it also means that there is no indication at all to go to approximate eigenstates; hence the spectroscopic picture (on which Bohr's scenario is based) gets lost.

Another option is to consider different points of view quasi simultaneously:

- classical nuclei, if reaction paths are of primary interest; in this approach, energy dissipation into translatory degrees of freedom (at least for a reaction of just two molecules) can relatively easily be encorporated, because there is no entanglement between internal and translatory degrees of freedom;

- (approximate) eigenstates, if spectroscopic behaviour is of primary interest.

These points of view are "complementary", and using them both at the same time needs a feeling or flair for molecular behaviour, which chemists try to learn by experience.

Incidentally, the Bohr scenario introduces a *stochastic* dynamics, as it gives (via Fermi Golden Rule, for example) transition probabilities between different energy eigenstates. The time-dependent Schrödinger equation, taken for itself, does *not* contain any stochastic aspects; it is completely regular and smooth (differentiable), without any sort of quantum jumps between energy eigenstates.

Also, the transition

$$\psi_C \bar{\otimes} \psi_D \quad \overset{?}{\rightarrow} \quad \psi_{C,m} \bar{\otimes} \psi_{D,n} \tag{9}$$

in eq. (8) is a kind of dissipative effect: The Schrödinger equation for itself will not transform a non-eigenstate into an eigenstate. This sort of transformation is usually built in by help of the projection postulate in connection with a measurement-type process: "Measurement" of an observable $\hat{A}$ like the energy of a system is said to transform the state vector ψ in question into an eigenstate $\tilde{\psi}$ of $\hat{A}$,

$$\hat{A}\tilde{\psi} = \tilde{a}\tilde{\psi}. \tag{10}$$

This brings again in a stochastic aspect, since different eigenstates $\tilde{\psi}$ may arise (starting from the same initial state ψ) with certain probabilities, in such a way that – on the average – the original expectation value $\langle \psi | \hat{A} \psi \rangle$ is conserved.

One may also replace the family $\psi_{C,m} \bar{\otimes} \psi_{D,n}$ of (product species) energy eigenstates by another family Φ_v, $v = 1, 2, \ldots$ of "preferred" states and compute transition probabilities

$$\left| \langle (\psi_{A,j} \otimes \psi_{B,k})_t | \Phi_v \rangle \right|^2 \tag{11}$$

and the corresponding *conditional* reaction rates

$$k_v \propto \frac{d}{dt} \left| \langle (\psi_{A,j} \otimes \psi_{B,k})_t | \Phi_v \rangle \right|^2. \tag{12}$$

These reaction rates are called *conditional* since they make sense only under the condition that the final states of the time-evolution are indeed members of the family Φ_v, $v = 1, 2, \ldots$. Otherwise the rates of eq. (12) are senseless.

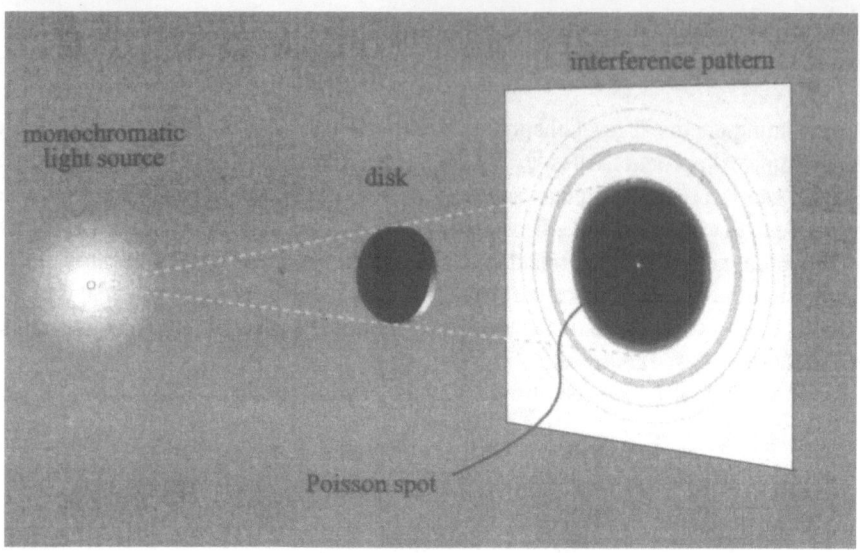

Figure 1. Interference pattern of a monochromatic light source diffracted at a circular disk. It shows the famous Poisson spot in the center of the geometric shadow.

2. IS IT WORTH RETHINKING QUANTUM-MECHANICAL THEORY UNDERLYING CHEMISTRY?

Let us go back for a moment to the year 1818, when Augustin Fresnel presented his memoir on diffraction for the prize of the French Academy. At the time, academies still played a vital role in scientific life. Poisson, Biot, and Laplace were on the committee to judge the paper, and all three were declared supporters of the corpuscular theory of light. Fresnel had calculated various cases of his formula and compared them with experiment. But Poisson noticed that one could calculate the integral for a circular disk, and came to the conclusion that there is a white spot in the center of the shadow of a circular disk. He raised this as an objection to the whole theory, since it was patently ridiculous. However, Arago and Fresnel went and performed the experiment, and, sure enough the spot was there, whereupon Fresnel was awarded the prize. This phenomenon is now known as "Poisson spot". [1]

Hence results exist in the natural sciences, which do not conform everyday intuition. Results which need a long time to be raised to technical application. In case of wave theory of light such applications are interferometry and X-ray structure analysis, for example, and both were developed many years after Fresnel's death.

What we can learn from this story is, that "strange" aspects of a theory should periodically be checked again. Curious results should not be overlooked and ignored, but kept in mind for later and better understanding.

Quantum mechanics is in some respects a strange theory in the above sense:

- Superpositions of certain states are counterintuitive, particularly when we think of macroscopic systems which are often expected to behave exclusively according to the rules of classical physics. Most annoying are states which do not admit a dispersion-free unequivocal position (they can be constructed as superpositions of states with different dispersion-free positions).

- The measurement process is thought to go via instantaneous changes from a given state with state vector ψ_i to an eigenstate $\psi_{\hat{A}}$ of the observable $\hat{A}$ to be measured. Is this process really an instantaneous one, i. e., a quantum jump? Or is it possible to find a continuous (stochastic) dynamics which mimics quantum jumps in a suitable way?

Usually these strange aspects are ignored though they possibly could give rise to new technical applications and fundamental insight.

Interestingly enough, the strange aspects of quantum mechanics are not at all limited to macroscopic systems. Already in the molecular realm, similar problems arise. Chemists and molecular physicists, in particular, have difficulties with quantum mechanics, since the idea of a molecular frame or molecular structure is not more than just tolerated in quantum theory. There "molecular structure" plays a similar role as "position", with a dispersion-free molecular structure being more the exception than the rule. Nevertheless, the traditional chemical ideas of (dispersion-free) molecular structure are ingenious and fantastically well based on phenomenology. It is fascinating to see, how the whole diversity of organic chemistry can be treated with a few simple rules. Quantum mechanics alone would probably never have been able to give such simple high-quality guidelines as those worked out by chemists during the last 150 years.

Here we shall try to spot the "strange" aspects of quantum mechanics, and indicate how these aspects could be built into a general theory of *single* quantum systems. We shall mainly take molecules as examples, though other vehicles of discussion could equally well be used. It will turn out that many genuine quantum effects, and in particular, the interesting stochastic behaviour of quantum dynamics, averages out when only large ensemble of molecules are considered. To the contrary, single quantum systems[2-8] show features which are far from being trivially explained in terms of, either, traditional chemistry or standard quantum mechanics. Proper treatment of stochastic dynamics, external perturbations, behaviour under measurement-type processes and single-molecule spectroscopic experiments goes beyond the standard quantum-mechanical approach.

The basic idea here will be that stochastic pure-state time-evolution (based on the usual Schrödinger dynamics) is of central importance for the description of *single* quantum systems (as opposed to *ensembles* of quantum systems); further, that it is external influence on a given quantum system which gives rise to this stochastic pure-state dynamics. Though measurement-type dynamics will also play a role, interest is concentrated here on the question how a thermal density operator should be properly interpreted in terms of pure states: this problem is strongly connected with the different possible reaction paths starting in thermal equilibrium and also strongly connected with the detailed dynamical behaviour of complex systems such as chemical substances (going beyond gas phase at low pressure).

Quantum mechanics of single individual quantum objects is still in its infancy. It is, for example, not clear how a rigorous derivation of stochastic pure-state dynamics[9-14] could be achieved. This is not merely a mathematical problem: it is the conceptional background which is difficult to clarify.

For traditional *ensemble* spectroscopy, there is no necessity to introduce a stochastic dynamics: a (non-stochastic) density-operator dynamics (see below) does the job marvellously. Nevertheless even there one uses a stochastic language inspired by the Bohr scenario of (stochastic) quantum jumps between molecular energy eigenstates. Sometimes proper (symmetry-adapted) energy eigenstates are used, and sometimes only approximate energy eigenstates (corresponding to a sink in the Born–Oppenheimer hyperplane): hence this Bohr scenario as used in molecular spectroscopy is not clearly enough defined; it is mainly a language which leads quickly to interesting results and spectral predictions, without claiming to be the true picture for the description of single molecules. Here the Bohr scenario is taken as a starting point for more general stochastic pure-state dynamics.

3. FLUCTUATIONS IN QUANTUM MECHANICS

We consider $N = 10^{23}$ molecules of some chemical species at inverse temperature β and at low pressure, say 1mbar, such that the interaction between different molecules is relatively small. Every molecule j, $j = 1, 2, \ldots, 10^{23}$, of the ensemble (collection) considered is in a certain quantum-mechanical state which can be represented by a density operator D_j. Chemists would probably argue that the states D_j are pure, i. e., $D_j^2 = D_j$, but there is no fundamental reason for purity and this is also not essential in the following.

From a statistical point of view, one could look, for example, at the *average state*

$$\bar{D} \stackrel{\text{def}}{=} \frac{1}{N} \sum_{j=1}^{N} D_j, \tag{13}$$

In a thermal situation this average state $\bar{D}$ would coincide with a thermal density operator, such as

$$D_\beta = \frac{\exp\{-\beta H\}}{\text{Trace}(\exp\{-\beta H\})} = \frac{1}{\sum\limits_{j=1}^{N} e^{-\beta E_j}} \begin{pmatrix} e^{-\beta E_1} & 0 & \cdots & 0 \\ 0 & e^{-\beta E_2} & & \vdots \\ \vdots & & \ddots & 0 \\ 0 & \cdots & 0 & e^{-\beta E_N} \end{pmatrix}, \tag{14}$$

the latter expression referring to a particular basis of (strict or approximate) energy eigenfunctions. Apart from the average state, one might also compute and use the "variance"

$$\sigma_D^2 \stackrel{\text{def}}{=} \frac{1}{N-1} \sum_{j=1}^{N} (D_j - \bar{D})^* (D_j - \bar{D}) \tag{15}$$

describing the fluctuations around the average $\bar{D}$.

Interestingly enough, traditional quantum mechanics uses *only* the average state $\bar{D}$, but never the variance σ_D or any other statistical concept to incorporate the fluctuations of states D_j. This corresponds to the use of substances in chemistry instead of collections of single molecules with an interesting and intrinsic single-molecule dynamical behaviour.

The experimental techniques nowadays allow to look at single (or very few) molecules.[3,5,7,15-24] The interesting point is that single quantum objects show a stochastic behaviour which depends on the environment considered.

The difference between spectroscopy of an *ensemble* of molecules and spectroscopy of *single* molecules lies in the fact that

- in single-molecule spectroscopy spectral lines may change their absorption and emission frequencies in a stochastic way;

- in ensemble spectroscopy a spectral line does *not* change its absorption or emission frequency.

An example of stochastically varying (fluorescence emission spectroscopy) spectral lines is given in fig. 2. It refers to single terrylene defects in a hexadecane matrix. The spectrum of an *ensemble* of such terrylene defects would not change (in time), because the ensemble corresponds to an *average* over all different possible stochastic behaviours, and this average does not change (in time).

Another interesting aspect arises when we look at the matter the other way round, i. e., starting with the ensemble and proceeding to the single-molecule behaviour (this is the "inverse" of the average over all different stochastic spectral paths): The average (over an ensemble) in thermal equilibrium at inverse temperature β is described by a density-operator D_β.

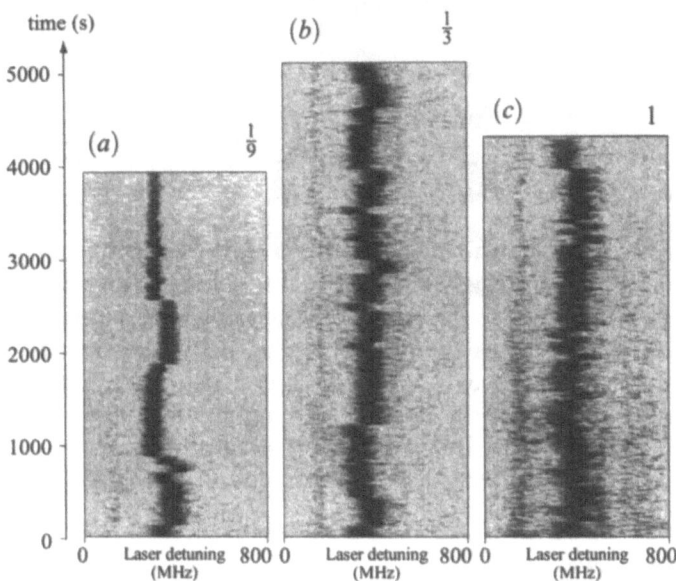

Figure 2. Migrating absorption frequencies attributed to one single terrylene defect in a hexadecane matrix. The shown migrating absorption lines were measured at different relative laser intensities $\frac{1}{9} : \frac{1}{3} : 1$ (total time almost 4 hours). In other examples of the same material the stochastic behaviour is even much more pronounced. The data were measured on February 18, 1994, by Mauro Croci, Viktor Palm, W. E. Moerner and Urs P. Wild. They can be downloaded from http://www.chem.ethz.ch/sms/. This figure is reproduced with permission of Prof. Urs Wild (ETH-Zürich).

Proceeding now to individual spectral paths can only be done by *decomposition* of the thermal density operator D_β into individual molecular states $\widetilde{D}_j$. Though the overall thermal state represented by the density operator D_β is stationary, the pure states $\widetilde{D}_j$ arising in the decomposition need not be stationary at all (this is shown by the example of single-molecule spectroscopy).

In our heuristic example above, we averaged over $N = 10^{23}$ different states D_j, ending up with the thermal state D_β. The corresponding decomposition of D_β into the (time-varying) states $\widetilde{D}_j$ is *not* unique, and hence it is not clear that a decomposition leads back to the states D_j used for averaging.

With single-molecule spectroscopy very interesting *stochastic* behaviours of the states D_j in an ensemble (collection) may arise, see fig. 2, since lines in the fluorescence spectrum of the (few) defect molecules begin to "hop around" with respect to the frequency (but without changing their position). Averaging over all sorts of stochastic behaviours of such defect molecules would lead back to the overall spectrum (encoded in the thermal density operator). If, for example, a particular line in the spectrum jumps to higher frequencies, then (on an average) this will be compensated by another line jumping to lower frequencies.

The averaging hence corresponds to the mixing of different pure states, ending up in one thermal density operator. And the other way round: If in single-molecule spectroscopy one looks at few defect molecules (by restricting the spatial field of investigation in the matrix to about 100μm × 100μm), this corresponds to a decomposition of the thermal density operator D_β into pure states (the particular defect molecule observed corresponding to one pure state which is part of the decomposition). Though D_β is stationary, the pure states (corresponding to defect molecules) need not be stationary, and indeed show the mentioned stochastic frequency-hopping.

Hence we have "two different sorts" of quantum mechanics:

- one for ensembles of molecules, described by a regular, differentiable density-operator dynamics

- and one for single molecules (or single quantum systems), described by a stochastic dynamics on the level of pure states

Averaging over the stochastic dynamics leads back to the regular density-operator dynamics. Decomposing the density-operator dynamics leads to a stochastic pure-state dynamics. Averaging is trivial, whereas decompositions are not unique and hence not canonically defined. One might, of course, hope to *derive* the stochastic behaviour on the pure-state level from the density-operator dynamics, but such a derivation has never been done in a mathematically rigorous way. Though discussing this point below, we might already say that it is *not clear at* all if the traditional quantum-mechanical formalism is capable of giving a proper description of the stochastic dynamics arising with single quantum systems.

4. DIFFERENT INTERPRETATIONS OF ENSEMBLE-SPECTROSCOPIC RESULTS

Ensemble spectroscopy of molecules depends *only* on the average $\bar{D} = D_\beta$ of the states D_j of the different molecules in the ensemble. Hence molecular parameters such as bond lengths and bond angles, which are determined by ensemble spectroscopy, are always *conditional*: *Under the condition* that NF_3 has a nuclear structure, the rotational spectrum $[\nu(^{14}NF_3) = 21.362\text{GHz}, \nu(^{15}NF_3) = 21.258\text{GHz}]$ implies an N-F-bond length of 1.37Å. If, on the other hand, NF_3 would not have a nuclear structure and hence not have an N-F-bond length, then 1.37Å would be a mere number characterizing in some way the rotational spectrum of NF_3.

An ensemble-spectroscopic experiment gives *only* the thermal density operator

$$
D_\beta = \frac{1}{\sum_{j=1}^{N} e^{-\beta E_j}}
\begin{pmatrix}
e^{-\beta E_1} & 0 & \cdots & 0 \\
0 & e^{-\beta E_2} & & \vdots \\
\vdots & & \ddots & 0 \\
0 & \cdots & 0 & e^{-\beta E_N}
\end{pmatrix}
\tag{16}
$$

but "nothing more."

The essential point behind this argument is that a non-pure state such as a thermal density operator D_β cannot be decomposed uniquely into (pure or nonpure) states D_j. One can therefore consider different scenarios to interpret ensemble-spectroscopic results which are based on different decompositions of the thermal density operator D_β.

We consider here two scenarios (out of many others), which can exemplify this point of view:

1) In *Bohr's scenario*, an atom or molecule is *always* in an *energy eigenstate*. *Under this condition* the average D_β can be shown to stem from

$$
D_\beta = \frac{e^{-\beta E_1}}{\sum_{j=1}^{N} e^{-\beta E_j}}
\underbrace{\begin{pmatrix}
1 & 0 & \cdots & 0 \\
0 & 0 & \cdots & 0 \\
\vdots & \vdots & & \vdots \\
0 & 0 & \cdots & 0
\end{pmatrix}}_{D_1 \hat{=} \Psi_1}
+ \cdots +
\frac{e^{-\beta E_N}}{\sum_{j=1}^{N} e^{-\beta E_j}}
\underbrace{\begin{pmatrix}
0 & \cdots & 0 & 0 \\
\vdots & & \vdots & \vdots \\
0 & \cdots & 0 & 0 \\
0 & \cdots & 0 & 1
\end{pmatrix}}_{D_N \hat{=} \Psi_N}
\tag{17}
$$

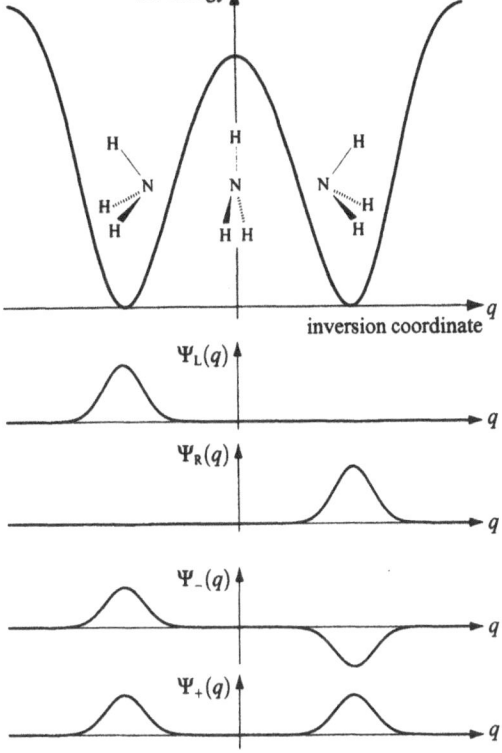

Figure 3. Illustration of the superposition principle using ammonia and ammonia-type molecules.

where the matrices are density operators $D_1, D_2, \ldots$, corresponding to energy eigenstate vectors

$$\hat{H}\Psi_k = E_k \Psi_k, \quad k = 1, 2, \ldots, N. \tag{18}$$

Furthermore the restriction of Bohr's scenario to eigenstates *implies* that transitions between different eigenstates Ψ_j and Ψ_k arise with probabilities proportional to $|\langle \Psi_j | \hat{B} \Psi_k \rangle|^2$, where $\hat{B}$ is the operator coupling the external radiation field (applied by the experimenter) to the molecules of the ensemble, as, for example, the dipole-moment operator.

2) The *chemist's scenario*, on the other hand, puts molecules preferentially into states admitting an (approximate or strict) *nuclear structure*. That this point of view is entirely different from Bohr's scenario becomes obvious when we look at ammonia-type molecules, see figs. 3 and 4. There we have left- and right-handed molecules Ψ_L, Ψ_R admitting a nuclear structure (corresponding to the chemist's scenario) as well as symmetry-adapted energy eigenstates Ψ_+ and Ψ_- without nuclear structure (corresponding to Bohr's scenario).

These two scenarios (out of many more), though being entirely different, are both compatible with ensemble-spectroscopic results, because the latter depend only on the thermal density operator D_β and because these different scenarios reflect only different decompositions of one and the same thermal density operator.

Incidentally, the state vectors Ψ_+ and Ψ_- in fig. 3 are the ground state and first excited state of ammonia (or ammonia-type species) and the transition between them is the famous

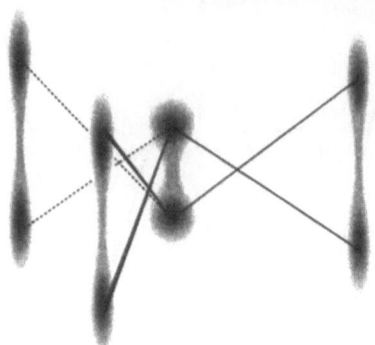

Figure 4. Sketched distribution of the nuclei in the ground state of an ammonia-type molecule.

MASER-transition. The mere existence of the MASER-transition suggests that the states Ψ_+ and Ψ_- do indeed exist, but this argument is not a rigorous and conclusive one.

Hence different interpretations of ensemble-spectroscopic results exist, corresponding to different decompositions of the thermal density operator D_β (one of these decompositions is given in eq. (17).

5. ENSEMBLES VS. SINGLE QUANTUM SYSTEMS

At this stage, we can summarize the differences between ensembles of quantum systems, on the one hand, and single quantum systems, on the other.

Ensembles of states D_j, $j = 1, 2, \ldots$, are described by their *average state*

$$\bar{D} \stackrel{\text{def}}{=} \frac{1}{N} \sum_{j=1}^{N} D_j. \tag{19}$$

An example of such an average state is given by a thermal state D_β. The respective dynamics $t \to \bar{D}(t)$ of average states

- is relatively simple (cf. the Bloch equations in NMR),

- is continuous and differentiable,

- does *not* make use of quantum jumps (à la Bohr),

- but does *not* give detailed information about the individual dynamics $t \to D_j(t)$.

The dynamics $t \to D_j(t)$ of *single* molecules (single quantum systems) shows stochastic (random) behaviour. *Different* single-molecule dynamics can give rise (by averaging) to one and the same average dynamics $t \to \bar{D}(t)$. Bohr's scenario gives no entirely satisfactory explanation of this stochastic behaviour. The experimental results from single-molecule spectroscopy, in particular, can only partly be explained by use of Bohr's scenario.

6. BOHR'S SCENARIO FOR SINGLE-MOLECULE SPECTROSCOPY

In Bohr's scenario, a defect molecule is considered as consisting of two (or more) energy levels Ψ_1 and Ψ_2, between which quantum jumps occur with certain transition probabilities.

The surrounding matrix molecules (numbered as $j = 1, 2, \ldots$) in states $\zeta_{j,1}$ or $\zeta_{j,2}$ exert an influence on the defect molecule in such a way that the level splitting ΔE of the defect molecule depends on the states of the surrounding matrix molecules (i. e., ΔE changes when a surrounding matrix molecule j changes from state $\zeta_{j,1}$ to $\zeta_{j,2}$). In other words: The stochastic change of ΔE and the corresponding stochastic behaviour in the fluorescence spectrum of the defect molecule is due to stochastic changes in the state structure of the surrounding matrix molecules which undergo quantum jumps between their two possible levels.

This classical, non-quantum-mechanical description is very useful. Instead of referring to the entire joint quantum system {defect molecule & surrounding matrix molecules}, the joint system is treated as split up into several parts, namely all the "molecules" building up the matrix, with a classical interaction, which influences the level splitting ΔE of one of them, namely the defect molecule. In such a scenario, the random behaviour of ΔE results from a random behaviour of the "classical environment" of the defect molecule.

The pure states used in this argument are thought to be (approximate) eigenstates. They are not energy eigenstates in a strict sense, because they refer to the isolated defect molecule or isolated matrix molecules. If the coupling between defect and matrix molecules would be considered, then these "eigenstates" would not be stationary with respect to the Schrödinger equation.

Alternatively, let us consider (symmetry-adapted) energy eigenstates Ψ_n of the *joint* system {defect molecule & surrounding matrix molecules}. In this description the eigenstates and associated energies do not change. Hence applying Bohr's description to these (strict) energy eigenstates will *not* give rise to changing frequencies or frequency-hopping in the spectrum. This immediately results in two problem areas:

- Even if one accepts quantum jumps between *strict* eigenstates, it is not clear how quantum jumps between *transient* states can be introduced (even if these transient states could be viewed as approximate eigenstates in some sense),

- different settings, one with strict eigenstates and one with approximate eigenstates, give rise to different results.

7. WHEN DOES BOHR'S SCENARIO BREAK DOWN?

Let us again have a look at the ammonia-type molecules. Though having analogous sets of pure states (such as Ψ_L, Ψ_R, Ψ_+ and Ψ_-), different ammonia-type molecules can have rather different level splitting $(E_- - E_+)$. In table 1 some ammonia-type species together with the respective (estimated) level splittings are listed. The heuristic idea[25] is that a large level splitting gives rise to quantum behaviour (all pure states arising), whereas a small level splitting gives rise to partly classical behaviour (only the handed states Ψ_L and Ψ_R arise).

This heuristic idea is based on chemical wisdom and spectroscopic expertise and has been discussed from many different viewpoints.[25,28-49] Currently we have no completely satisfactory picture for the transition from quantum to classical behaviour.

For our present purpose it is sufficient to consider the situation with respect to the following (hypothetical) assumptions:

- Let us, for the moment, assume that Bohr's scenario is roughly correct for *large* level splittings such as in monodeuteroaniline, ammonia and NHDT. This would mean that the energy eigenstates Ψ_+ and Ψ_- play a predominant role in spectroscopic situations.

- Let us further assume, that Bohr's scenario can also be adopted for the case of *small* level splitting, when for all practical purposes the wave functions arising are localized

Table 1. Ammonia-type molecules

	Barrier	Level Splitting
Monodeuteroaniline[27]	$5.5 \, kJ \, mol^{-1}$	$600 \, J \, mol^{-1}$
Ammonia	$23.9 \, kJ \, mol^{-1}$	$9.3 \, J \, mol^{-1}$
Naphthazarin[26]	$\approx 50 \, kJ \, mol^{-1}$	$\approx 0.02 \, J \, mol^{-1}$
Aspartic acid	$\approx 140 \, kJ \, mol^{-1}$	$\approx 10^{-60} \, J \, mol^{-1}$

in one of the two sinks of the Born–Oppenheimer potential. This is to say that localized approximate energy eigenfunctions (in one of the sinks) play a predominant role in spectroscopic situations.

- Assume, in addition, that our ammonia-type molecule is a defect molecule in a molecular matrix (as explained in section 3).

- Finally, assume that the level splitting can be changed, either by application of external fields or using defect molecules with different level splitting.

If Bohr's scenario holds true for large *and* small level splittings but with different eigenstates (delocalized and localized ones), there must be some domain of *intermediate* level splitting where neither class of eigenstates can be predominant. In this intermediate domain all sorts of pure states (such as Ψ_L, Ψ_R, Ψ_+ and Ψ_- and others) must arise and are expected to play an important role. Hence in this intermediate domain *the Bohr scenario is expected to brake down*.

Note that *we do not claim* that the Bohr scenario were really correct for very small or very large level splitting. In neither of these domains is it clear how to model the stochastic dynamics in order to get some realistic picture, and, in particular, the Bohr picture perhaps provides just a suitable language to speak about spectra. Our argument rather is of counterfactual flavour and could be reformulated in the following way: *If* Bohr's language can be successfully used to describe single-molecule behaviour for small and large level splitting, then it cannot be used for the particular domain of intermediate level splitting where strict, symmetry-adapted eigenstates die out, whereas localized approximate eigenstates begin to appear.

8. THE STOCHASTIC BEHAVIOUR OF SINGLE MOLECULES

The quantum theory of single, open quantum systems is still in its infancy and no entirely satisfactory and mathematically rigorous description has been given.

For a joint system {molecule & environment} one can usually not expect the subsystem consisting of the molecule alone to be in a pure state. This is the case only for product states of the joint system. Hence even if the energetic coupling between the molecule and its environment is negligible or zero, there is no rigorous mathematical reasoning leading to pure states of the molecule. Therefore we used density operators D_j in all the above arguments (density operators describe non-pure states).

Chemical heuristics makes extensive use of pure states. The strange point is: *If a molecule (or more general quantum system) coupled to a quantum environment is always in an exact product state, then no interesting facts arise, neither a stochastic dynamics nor quantum jumps nor chemical reactions*[50]. Nevertheless, to keep in touch with chemical heuristics, one must introduce *approximate* pure states and dynamics of approximate pure states for molecules[50].

Here only some account is given comparing Bohr's scenario to more general stochastic dynamics of pure states. Though using *pure* states here for convenience, we must always be aware that *approximate* pure states will appear in any mathematically and conceptually rigorous quantum theory of single quantum systems.

To compare Bohr's scenario with more general stochastic dynamics, we consider the proper ground state Ψ_+ of a chiral molecule. A chiral molecule is an ammonia-type species (cf. fig. 3), for which the state Ψ_+ has not yet been prepared experimentally. Although it is an exact energy eigenstate of the molecular hamiltonian, it is thought to be unstable under small external perturbations. It might, for example, decay into the handed states Ψ_L and Ψ_R.

Bohr's scenario insists on *instantaneous* transitions between different pure states. In our situation, these could be instantaneous transitions

$$\Psi_+ \xrightarrow{\text{probability } \frac{1}{2}} \Psi_L \tag{20}$$

$$\Psi_+ \xrightarrow{\text{probability } \frac{1}{2}} \Psi_R. \tag{21}$$

In a more general *quantum mechanics of single systems*, also *non-instantaneous, continuous* dynamics of pure states are considered[9, 10, 11, 13, 14, 51]. In our example, this means that the initial state Ψ_+ decays on a *continuous path* in the state space

- to Ψ_L with probability $P_L \approx \frac{1}{2}$

- to Ψ_R with probability $P_R \approx \frac{1}{2}$.

Possible continuous random paths of pure states are sketched in fig. 5. Actually, paths of pure state *vectors* (i. e., wave vectors) are given in fig. 5 and not pure states themselves. Depending on the state of the environment (denoted by χ_1 and χ_2), the initial state Ψ_+ decays into different final states. The time-evolved states are denoted by $(\psi_+)_{t,\chi_1}$ and $(\psi_+)_{t,\chi_2}$, respectively. In case of Bohr's scenario, the intermediate state (vectors) in fig. 5 would not appear at all.

Some attempts have been made to *derive* stochastic pure-state dynamics[9, 10, 11, 13, 14, 51]. If one takes as starting point the joint system {quantum object & quantum environment}, then different coupling terms in the joint hamiltonian can lead to different pure-state dynamics of the quantum object in question. The expectation-value postulate, for example, may be fulfilled with respect to one particular choice of coupling constants, but not in general. Therefore, one may get transition probabilities

$$P_L := P_{+,L} < \frac{1}{2}, \qquad P_R := P_{+,R} > \frac{1}{2} \tag{22}$$

and vice versa, instead of exact probabilities of $\frac{1}{2}$[52]. Though these derivations are not perfect from a conceptual point of view, they open up various interesting options. If the expectation value postulate need not be exactly fulfilled, one may go even a step further and ask if the state of the environment, denoted by χ in fig. 5, can be controlled in a clever way such that, for example, the state Ψ_+ decays with 95% probability into Ψ_L (instead of probability $\frac{1}{2}$) and with

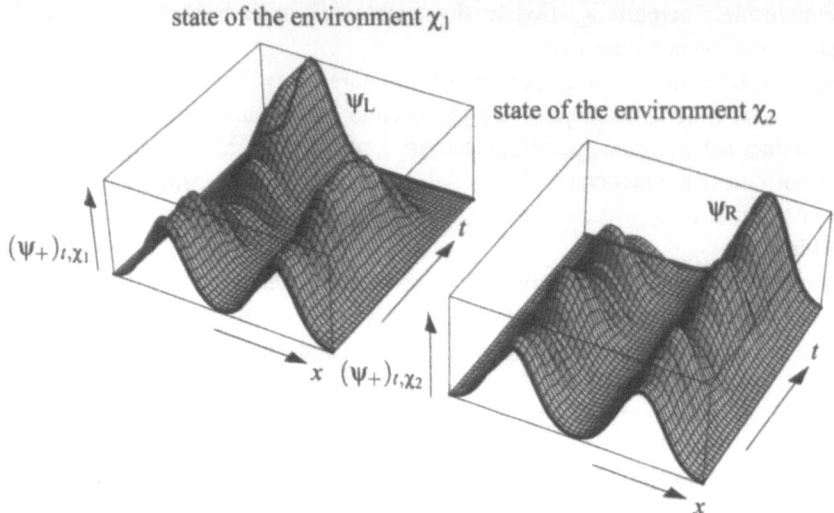

state of the environment χ_1

ψ_L

state of the environment χ_2

$(\psi_+)_{t,\chi_1}$

ψ_R

t

x $(\psi_+)_{t,\chi_2}$

t

x

Figure 5. Sketch of a *continuous* stochastic dynamics with initial state ψ_+ and different final states ψ_L and ψ_R (of the dressed quantum object), respectively. The stochastic dynamics $(\psi_+)_{t,\chi}$ depends on the time t and the state χ of the environment, ending up either with the final state ψ_L or the final state ψ_R. In the Bohr scenario, *quantum jumps* are taken instead, i. e., instantaneous transitions $\psi_+ \to \psi_L$ and $\psi_+ \to \psi_R$ with transition probabilities $P_L := P_{+,L}$ and $P_R := P_{+,R}$. The expectation value postulate claims $P_{+,L} = P_{+,R}$.

5% probability to Ψ_R (instead of probability $\frac{1}{2}$). This would mean that one could "deceive" the expectation-value postulate and open up reaction channels in a selective way.

In classical dynamical systems, similar attempts have been made and run under the name of *stochastic control* [53, 54, 55]: the more random or chaotic a stochastic system behaves, the more it is accessible to stochastic control, and any chosen state can be reached by infinitely small perturbation of an external parameter.

9. CHEMICAL KINETICS AS A QUANTUM-MECHANICAL THEORY

In section 1, we introduced conditional reaction rates

$$k_\nu \propto \frac{d}{dt} \left| \langle (\psi_{A,j} \otimes \psi_{B,k})_t \,|\, \Phi_\nu \rangle \right|^2 . \qquad (23)$$

As already mentioned, in traditional quantum mechanics there is no reason whatsoever to end up with particular states Φ_ν (such as energy eigenstates or states admitting a nuclear structure, and so on). Therefore, a particular family Φ_ν, $\nu = 1, 2, \ldots$ of states is usually introduced "by hand", guided by intuition, experimental results or just as a kind of boundary condition which allows an easy conceptual understanding.

In a general *quantum theory of single systems*, one would try to derive an appropriate stochastic dynamics of approximate pure states, depending on the state of the quantum environment. The correct family Φ_ν, $\nu = 1, 2, \ldots$ of final states would then automatically emerge from the stochastic dynamics itself. In such a derivation one cannot expect that the states Φ_ν, $\nu = 1, 2, \ldots$ arise exclusively but must expect that all possible states arise (as final states), though with largely differing probabilities (highest probabilities being attributed to the family of states Φ_ν).

Hence from the point of view of chemical kinetics, a quantum theory of single systems should answer the following questions: Given a a reaction

$$A+B=C+D, \qquad (24)$$

what are the final states? Product states with respect to the species C and D? Energy eigenstates? States admitting a nuclear structure? Where does dissipation come in? Can dissipation be treated without ad hoc assumptions as expressed in eq. (23)? How does one get a preferred family Φ_v of states? What is the connection between reactions of a few molecules (of species A, B, C, and D) with reactions of substances (of the same species)? How can substances be properly defined in quantum mechanics, such that concentrations c_A, c_B, c_C, c_D or chemical potentials μ_A, μ_B, μ_C, μ_D are (approximate) classical observables?[56-59] Can one select reaction channels (leading to one particular final state Φ_{v_o}) on a detailed level by external stochastic control, i. e., by control of the state χ of the environment (as in fig. 5)? How chaotic (or how ergodic, or how mixing) is a chemical reaction system? Is it necessary to use quantum jumps? Or can a more detailed (stochastic but) continuous time-evolution replace and mimic quantum jumps in a chemical reaction system?

10. ACKNOWLEDGMENT

The author gratefully acknowledges the preparation of the figures in this paper by Dr. Pitt Funck (H&Q Zürich).

REFERENCES

1. V. Guillemin and S. Sternberg. "Symplectic Techniques in Physics," Cambridge University Press, Cambridge (1984).
2. L. Fleury, A. Zumbusch, M. Orrit, R. Brown and J. Bernard, *J. Luminescence* 56:15 (1993).
3. W. E. Moerner, T. Plakhotnik, T. Irngartinger, M. Croci, V. Palm and U. P. Wild, *J. Phys. Chem.* 98:7382 (1994).
4. M. Pirotta, F. Güttler, H.-R. Gygax, A. Renn, J. Sepiol and U. P. Wild, *Chem. Phys. Lett.* 208:379 (1993).
5. U. P. Wild, M. Croci, F. Güttler, M. Pirotta and A. Renn, *J. Luminescence* 60 & 61:1003 (1994).
6. G. Zumofen and J. Klafter, *Chem. Phys. Lett.* 219:303 (1994).
7. M. Croci, H.-J. Müschenborn, F. Güttler, A. Renn and U. P. Wild, *Chem. Phys. Lett.* 212:71 (1993).
8. W. P. Ambrose and W. E. Moerner, *Nature* 349:225 (1991).
9. H. Primas, The measurement process in the individual interpretation of quantum mechanics, *in*: "Quantum Theory without Reduction," M. Cini and J.-M. Lévy-Leblond, ed., IOP Publishing Ltd., Bristol (1990).
10. H. Primas, Induced nonlinear time evolution of open quantum objects, *in*: "Sixty-two Years of Uncertainty: Historical, Philosophical, and Physical Inquiries into the Foundations of Quantum Mechanics," A. I. Miller, ed., Plenum Press, New York (1990).
11. N. Gisin, *Phys. Rev. Lett.* 52:1657 (1984).
12. N. Gisin, *Helv. Phys. Acta* 62:363 (1989).
13. G. C. Ghirardi, A. Rimini and T. Weber, *Phys. Rev. D* 34:479 (1986).
14. G. C. Ghirardi, P. Pearle and T. Weber, *Phys. Rev. A* 42:78 (1990).
15. J. Wrachtrup, "Magnetische Resonanz an einzelnen Molekülen und kohärente ODMR-Spektroskopie an molekularen Aggregaten in Festkörpern," Thesis FU Berlin, Berlin (1994).
16. T. Plakhotnik, W. E. Moerner, V. Palm and U. P. Wild, *Opt. Comm.* 114:83 (1995).
17. W. E. Moerner and T. Basché, *Angew. Chem. Int. Ed.* 32:457 (1993).
18. F. Güttler, J. Sepiol, T. Plakhotnik, A. Mitterdorfer, A. Renn and U. P. Wild, *J. Luminescence* 56:29 (1993).
19. W. E. Moerner, *Science* 265:46 (1994).
20. U. P. Wild, F. Güttler, M. Pirotta and A. Renn, *Chem. Phys. Lett.* 193:451 (1992).
21. F. Güttler, T. Irngartinger, T. Plakhotnik, A. Renn and U. P. Wild, *Chem. Phys. Lett.* 217:393 (1994).
22. T. Basché, W. E. Moerner, M. Orrit and U. P. Wild, "Single-Molecule Optical Detection, Imaging and Spectroscopy," VCH, Weinheim (1996).
23. H. Bach, A. Renn and U. P. Wild, *Chem. Phys. Lett.* 266:317 (1996).

24. T. Irngartinger, H. Bach, A. Renn and U. P. Wild, Fluorescence microscopy of single molecules: temperature dependence of linewidths, *in*: "Electric and Related Properties of Organic Solids," R. W. Munn, ed., Kluwer, Dordrecht (1997).

25. P. Pfeifer. "Chiral Molecules – a Superselection Rule Induced by the Radiation Field," Thesis ETH-Zürich No. 6551, ok Gotthard S+D AG, Zürich (1980).

26. J. R. d. l. Vega, J. H. Busch, J. H. Schauble, K. L. Kunze and B. E. Haggert, *J. Am. Chem. Soc.* 104:3295 (1982).

27. M. Quack, *Jahrbuch der Akademie der Wissenschaften zu Berlin 1990 - 1992* (1993).

28. M. Quack, *Chem. Phys. Lett.* 132:147 (1986).

29. M. Quack, *Angew. Chem. Int. Ed. Engl.* 28:571 (1989).

30. H. Spohn and R. Dümcke, *J. Stat. Phys.* 41:389 (1985).

31. H. Spohn, *Commun. Math. Phys.* 123:277 (1989).

32. A. Amann, *Synthese* 97:125 (1993).

33. A. Amann, *J. Chem. Phys.* 96:1317 (1992).

34. A. Amann, *Ann. Phys.* 208:414 (1991).

35. L. Arnold. "Stochastic Differential Equations: Theory and Applications," Interscience, New York (1974).

36. J. A. Cina and R. A. Harris, *J. Chem. Phys.* 100:2531 (1994).

37. J. A. Cina and R. A. Harris, *Science* 267:832 (1995).

38. P. Claverie and S. Diner, *Israel J. Chem.* 19:54 (1980).

39. P. Claverie and G. Jona-Lasinio, *Phys. Rev. A* 33:2245 (1986).

40. G. Jona-Lasinio and P. Claverie, *Progr. Theor. Phys. Suppl.* 86:54 (1986).

41. G. Jona-Lasinio, F. Martinelli and E. Scoppola, *Commun. Math. Phys.* 80:223 (1981).

42. B. T. Sutcliffe, The concept of molecular structure, *in*: "Theoretical Models of Chemical Bonding, Part 1: Atomic Hypothesis and the Concept of Molecular Structure," Z. B. Maksić, ed., Springer, Berlin (1990).

43. B. T. Sutcliffe, *J. Mol. Struct. (Theochem)* 259:29 (1992).

44. S. J. Weininger, *J. Chem. Ed.* 61:939 (1984).

45. A. S. Wightman and N. Glance, *Nucl. Phys. B (Proc. Suppl.)* 6:202 (1989).

46. R. G. Woolley, *Chem. Phys. Lett.* 125:200 (1986).

47. R. G. Woolley, *New Scientist 22 October 1988* 53 (1988).

48. R. G. Woolley, Quantum theory and the molecular hypothesis, *in*: "Molecules in Physics, Chemistry and Biology. Vol. 1," J. Maruani, ed., Kluwer, Dordrecht (1988).

49. R. G. Woolley, *J. Mol. Struct. (Theochem)* 230:17 (1991).

50. A. Amann and H. Atmanspacher, "Fluctuations in the Dynamics of Single Quantum Systems", preprint (1997).

51. H. Primas, Mathematical and philosophical questions in the theory of open and macroscopic quantum systems, *in*: "Sixty-two Years of Uncertainty: Historical, Philosophical, and Physical Inquiries into the Foundations of Quantum Mechanics," A. I. Miller, ed., Plenum Press, New York (1990).

52. A. Amann, *J. Math. Chem.* 18:247 (1995).

53. F. Moss and K. Wiesenfeld, *Scientific American* August 1995:50 (1995).

54. T. Shinbrot, C. Grebogi, E. Ott and J. A. Yorke, *Nature* 363:411 (1993).

55. K. Wiesenfeld and F. Moss, *Nature* 373:33 (1995).

56. U. Müller-Herold, *Lett. Math. Phys.* 4:45 (1980).

57. U. Müller-Herold, *Fortschr. Physik* 30:1 (1982).

58. U. Müller-Herold, *Lett. Math. Phys.* 8:127 (1984).

59. U. Müller-Herold, *J. Chem. Ed.* 62:379 (1985).

THE MOLECULE AND ITS ENVIRONMENT

B. T. Sutcliffe

Department of Chemistry
University of York
York YO1 5DD, England

1. INTRODUCTION

When we speak of molecules in an environment we are sometimes thinking about an isolated molecule in an environment provided by an electromagnetic field and sometimes we are thinking about the molecule in a material environment, for example in a dilute gas of other molecules or as a part of a liquid or as a solute molecule in a bulk solvent or as part of a molecular crystal. We hope to be able to understand molecular properties in an environment in terms of the interactions between molecules and, if relevant, the electromagnetic field. It is from this perspective that intermolecular interactions will be considered in what follows.

The notion of field and matter as separate entities is rather difficult to sustain within the theory of relativity, so the idea of the field as an environment for a molecule is somewhat problematic in modern relativity theory. The idea on an isolated molecule in contradistinction from its material environment, is problematic within pioneer quantum mechanics, since a single equation describes both the molecule and the environment. Because there is no relativistic quantum theory of electrons and nuclei together, it is not possible to deal with both these groups of problems simultaneously in an adequate way. It is the custom in dealing with chemical phenomena however, to assume that pioneer quantum mechanics is the correct theory to use and to treat field and hence relativistic effects as perturbations. This is the line that we shall take. But the problem that we need to discuss first is that of identifying the isolated molecule and we shall begin by explaining why it is tricky to do so. Indeed the idea of a molecule was a tricky one, even in classical chemistry long before the advent of quantum mechanics and for an introduction to some aspects of the puzzles involved and to the relevant literature, the reader may care to consult Sutcliffe[1].

Pioneer Quantum Mechanics

From a quantum mechanical perspective, the idea of a molecule in an environment provided by other molecules, is one that is not quite obvious. Both the notion of a *molecule* and that of an *environment* are classical ideas and have to be realised within quantum mechanics according to particular conventions. If we assume that the ordinary Schrödinger equation is sufficient, at least to first order, to describe the systems of interest to us, then we can specify the time-independent Schrödinger Hamiltonian describing a system of N charged particles in a coordinate frame fixed in the laboratory as

$$\hat{H}(\mathbf{x}) = -\frac{\hbar^2}{2} \sum_{i=1}^{N} m_i^{-1} \nabla^2(\mathbf{x}_i) + \frac{e^2}{8\pi\varepsilon_o} \sum_{i,j=1}^{N} {}'\frac{Z_i Z_j}{x_{ij}} \qquad (1)$$

where the separation between particles is defined by

$$x_{ij}^2 = \sum_\alpha (x_{\alpha j} - x_{\alpha i})^2 \qquad (2)$$

It is convenient to regard $\mathbf{x}_i$ as a column matrix of three cartesian components $x_{\alpha i}$, $\alpha = x, y, z$ and to regard $\mathbf{x}_i$ collectively as the 3 by N matrix $\mathbf{x}$. Each of the particles has mass m_i and charge $Z_i e$. The charge-numbers Z_i are positive for a nucleus and minus one for an electron. In a neutral system the charge-numbers sum to zero. The Schrödinger equation for the (constant energy) system is then

$$\hat{H}(\mathbf{x})\psi(\mathbf{x}) = E\psi(\mathbf{x}) \qquad (3)$$

This form is appropriate for any number of particles and there is nothing in it that would give us any clue about which bits of the operator correspond to molecules and which to environment. The form might also represent a chemical reaction at constant energy and it is equally uninformative about which bits are the products or the reactants. Or the form might represent an isolated neutral molecule but if it did, it could equally well represent all the products into which the molecule could dissociate while the system remains electrically neutral. We could go on: the form could represent a solid or a liquid, the number of particles could be Avogadro's number; some thinkers suppose that the appropriate value for N is always Eddington's number, the total number of particles in the universe. Difficulties of this kind make it extremely hard to talk in an unambiguous way in a quantum mechanical context of such apparently uncontroversial ideas as molecular or chemical structure. There is, in consequence, a quite extensive literature on these ambiguities and some aspects of it are treated in the book arising from the first *Small Molecules Indaba*.[2]

It is clear that if we are to understand the outcomes of experiments in terms of quantum mechanics, we must make choices and judgements about how to divide the equation up to produce the most tractable and informative way possible of dealing with the problem and also to yield results that make sense of our experience. But it should always be kept in mind that we are making choices and judgements, admittedly ones which are heavily circumscribed by experiment, but about which, nevertheless, there can be honest and sensibly argued differences of opinion.

To begin with we shall summarise what can be said on a purely general basis, of the solutions to the problem posed in terms of the pioneer quantum mechanical Hamiltonian. Then we shall show how the problem can be reformulated to yield solutions that may plausibly identified with an isolated molecule. The use of these solutions in describing a molecule in an environment provided by other molecules will then be discussed. Then we shall turn to the isolated molecule in a static electromagnetic field and try to see what we might hope for from solutions to the problem posed in this kind of environment.

The view that we shall take is *extrinsic* to the quantum mechanical problem as such and will involve clothing a particular model of experience in quantum mechanics. Of course not any old model will do: it has to be one that leads to results that are congruent with our experience, so the procedure is not a mere matter of taste. However more than one model may fill the bill and it is very improbable that one model alone will suffice to account for all of our experience of a particular type.

2. THE FREE SYSTEM

The Intrinsic Symmetry Aspects of the Problem

The Hamiltonian (1) is invariant under uniform translations in the frame fixed in the laboratory. This means that the centre of molecular mass moves through space like a free particle and the states of a free particle are not quantised and eigenfunctions are not square integrable. The centre of mass motion must therefore be separated out to disentangle any bound states from the translational continuum.

The Hamiltonian is also invariant under all orthogonal transformations (rotation-reflections) of the particle variables in the frame fixed in the laboratory. It is therefore possible to separate, at least to some extent, the orientational motions of the system from its purely internal motions. The internal motions comprise dilations, contractions and deformations of a specified configuration of particle variables. Put colloquially, the internal motions simply change the system geometry. But in any case solutions of the equation must form basis functions for irreducible representations of the orthogonal group in three dimensions, $O(3)$.

Permutation of the variable sets of identical particles also leaves the Hamiltonian invariant. So the solutions of the equation must form a basis for the permutation group of the system. The permutation group of the system is formed as the direct product of the permutation groups for each of the sets of identical particles. The fact that the particles involved are either fermions or bosons restricts the kind of representations that can arise.

If we stick to the time-independent Schrödinger equation in the absence of external fields or other perturbations, then this is all the symmetry that there is. Subsequently we shall see that in certain contexts these symmetries are either lowered or even broken. An example is the point group symmetry that arises naturally in accounting for the vibration-rotation spectra of isolated molecules. That this happens must always be a consequence of the particular model that is being treated quantum mechanically and the account so provided is appropriate only in so far as the model yields results in line with experiment.

Whether or not the Schrödinger equation has any bound state solutions even when the translational continuum has been removed, is actually a rather hard question to answer in general. An outline of what is at present known together with some references can be found in Sutcliffe.[3] Of course this emphasis on bound states should not obscure the fact that even using the translationally invariant Hamiltonian, there will be among its solutions, continuum states resulting from the relative motion of two or more fragments. The energies of the continuum states will be above the energies of the bound states (if any) and thus will not give rise to the problems that the translation continuum does. Such states can be treated, formally at least, in an expansion by stipulating that the sum includes integration over the continuum. In future expansions this interpretation will be assumed where appropriate but in practice, of course, such integration cannot be achieved. These continuum states are of the greatest interest in any discussions of scattering and hence of reactions.

The separation involves an essentially arbitrary choice of translationally invariant coordinates and there is always one less such coordinate than the original number because of the centre of mass coordinate. Thus after this separation is made, it is a matter of opinion and/or convention how the translationally invariant coordinates should be identified. So the role of the coordinates in specifying either electronic or nuclear motions becomes problematic. Of course the exact solutions are indifferent to the translationally invariant coordinate choice made and a set can always chosen that is convenient for the purposes in hand. Since our aim is to identify a molecule we shall begin by distinguishing between electronic and nuclear coordinates in the laboratory frame. To distinguish between electrons and nuclei, the variables are split up into two sets, one set consisting of L variables, $\mathbf{x}_i^e$, describing the electrons and the other set of H variables, $\mathbf{x}_i^n$, describing the nuclei and $N = L + H$. When it is necessary to emphasise this split, (1) will be denoted $\hat{H}(\mathbf{x}^n, \mathbf{x}^e)$.

For a system with more than two particles one can transform the coordinates such that rotational motion can be expressed in terms of three translationally invariant orientation variables, with the remaining motions expressed in terms of variables (commonly called internal coordinates) which are invariant under translations and all orthogonal transformations. The coordinate transformation is often referred to as constructing a frame fixed in the body and it is supposed that the three orientation variables are specified by means of an orthogonal matrix $\mathbf{C}$, the elements of which are expressed as functions of three eulerian angles $\phi_m, m = 1,2,3$ which are orientation variables and so there will be just $3N - 6$ internal variables for the particles. The matrix $\mathbf{C}$ relates a set of cartesian coordinates $\mathbf{z}$ in the frame fixed in the body to translationally invariant combination of the cartesian coordinates fixed in the laboratory as

$$\mathbf{x} \rightarrow \mathbf{C}\mathbf{z} \tag{4}$$

Since $\mathbf{z}$ are fixed in the body, not all their $3N - 3$ components are independent, for there must be three relations between them. Hence components of $\mathbf{z}_i$ must be writable in terms of $3N - 6$ independent internal coordinates $q_i, i = 1,2,\ldots,3N-6$. Some of the q_i maybe components of $\mathbf{z}_i$ but generally q_i are expressible in terms of scalar products of the $\mathbf{x}_i$ (and equally of the $\mathbf{z}_i$) since scalar products are the most general constructions that are invariant under orthogonal transformations of their constituent vectors.

As an example, if it is assumed that the ϕ_m, $m = 1,2,3$ can be written as functions of the $\mathbf{x}_i^n$ alone and that $3H - 6$ of the q_i can also be so written then it can be shown that

$$\frac{\partial}{\partial \mathbf{x}_i^n} = m_i M_T^{-1} \frac{\partial}{\partial \mathbf{X}_T} + \mathbf{C}(\frac{\partial}{\partial \mathbf{n}_i} - m_i M^{-1} \sum_{j=1}^{L} \frac{\partial}{\partial \mathbf{z}_j}) \tag{5}$$

where

$$\frac{\partial}{\partial \mathbf{n}_i} \equiv \mathbf{\Omega}^i \mathbf{D} \frac{\partial}{\partial \phi} + \mathbf{Q}^i \frac{\partial}{\partial \mathbf{q}} + \mathbf{\Omega}^i \sum_{j=1}^{L} \hat{z}_j \frac{\partial}{\partial \mathbf{z}_j}$$

The L "electronic" coordinates $\mathbf{z}_i$ are transformed as in (4) as new variables relative to the centre of nuclear mass. With this choice it also follows that

$$\frac{\partial}{\partial \mathbf{x}_i^e} = m M_T^{-1} \frac{\partial}{\partial \mathbf{X}_T} + \mathbf{C} \frac{\partial}{\partial \mathbf{z}_i} \tag{6}$$

M_T is the total mass of the system, M the mass of the nuclei and $\mathbf{X}_T$ the centre of mass coordinate. The column matrix of the cartesian components of the partial derivative operator is written as $\partial/\partial \mathbf{x}_i$ and when it seems more convenient this column matrix of derivative operators will also be denoted as the vector grad operator $\vec{\nabla}(\mathbf{x}_i)$. The skew-symmetric matrix $\hat{z}_i$ is

$$\hat{z}_i = \begin{pmatrix} 0 & -z_{zi} & z_{yi} \\ z_{zi} & 0 & -z_{xi} \\ -z_{yi} & z_{xi} & 0 \end{pmatrix} \tag{7}$$

or

$$\hat{z}_i = \sum_{\alpha} z_{\alpha i} \mathbf{M}^{\alpha T}$$

where the $\mathbf{M}^{\alpha}$ are the standard infinitesimal rotation generators. A variable symbol with a caret over it will, from now on, be used to denote a skew-symmetric matrix as defined by (7).

Clearly the last term in (5) could be written as

$$\sum_{j=1}^{L} \mathbf{z}_j \times \vec{V}(\mathbf{z}_j)$$

where $\times$ denotes the vector product.

The 3 by 3 matrix $\mathbf{\Omega}^i$ and and the 3 by $3H-6$ matrix $\mathbf{Q}^i$ both have elements which are functions only of the q_k while the 3 by 3 matrix $\mathbf{D}$ is composed of trigonometric functions of the ϕ_m only.

The transformation of the Schrödinger Hamiltonian to these new coordinates is rather involved and we shall simply quote the results that we need. The complete kinetic energy operator may be written as

$$\hat{K}(\mathbf{z}) + \hat{K}(\mathbf{q}, \mathbf{z}) + \hat{K}(\phi, \mathbf{q}, \mathbf{z}) \tag{8}$$

The first term in (8) represents the electronic kinetic energy

$$\hat{K}(\mathbf{z}) = -\frac{\hbar^2}{2\mu} \sum_{i=1}^{L} \nabla^2(\mathbf{z}_i) - \frac{\hbar^2}{2M} \sum_{i,j=1}^{L}{}' \vec{\nabla}(\mathbf{z}_i) \cdot \vec{\nabla}(\mathbf{z}_j) \tag{9}$$

while the last term is

$$\hat{K}(\phi, \mathbf{q}, \mathbf{z}) = \frac{1}{2}\left(\sum_{\alpha\beta} \kappa_{\alpha\beta} \hat{L}_\alpha \hat{L}_\beta + \hbar \sum_\alpha \bar{\lambda}_\alpha \hat{L}_\alpha\right) \tag{10}$$

Here the angular momentum expressed in the frame fixed in the body is

$$\hat{\mathbf{L}}(\phi) = \frac{\hbar}{i} \mathbf{D} \frac{\partial}{\partial \phi} \tag{11}$$

The components of this operator obey the *standard* commutation conditions. The matrix κ is an inverse generalised inertia tensor defined as the 3 by 3 matrix

$$\kappa = \sum_{i}^{H} m_i^{-1} \mathbf{\Omega}^{i^T} \mathbf{\Omega}^i \tag{12}$$

and

$$\bar{\lambda}_\alpha = (\lambda_\alpha + 2(\kappa \hat{\mathbf{i}})_\alpha) \tag{13}$$

with $\hat{\mathbf{i}}$ as a 3 by 1 column matrix of cartesian components

$$\hat{\mathbf{i}} = \frac{1}{i} \sum_{i=1}^{L} \hat{\mathbf{z}}_i \frac{\partial}{\partial \mathbf{z}_i} \tag{14}$$

and

$$\lambda_\alpha = \frac{1}{i}\left(\nabla_\alpha + 2 \sum_{k=1}^{3H-6} \tau_{k\alpha} \frac{\partial}{\partial q_k}\right) \tag{15}$$

with the $3H-6$ by 3 matrix τ defined as

$$\tau = \sum_{i}^{H} m_i^{-1} \mathbf{Q}^{i^T} \mathbf{\Omega}^i \tag{16}$$

29

and

$$V_\alpha = \sum_{i=1}^{H} m_i^{-1} \left(\sum_\beta ({\Omega^i}^T M^\beta \Omega^i)_{\beta\alpha} + \sum_{l=1}^{3H-6} \left({Q^i}^T \frac{\partial}{\partial q_l} \Omega^i \right)_{l\alpha} \right) \tag{17}$$

The term (15) is associated with the Coriolis coupling and so no coordinate system can be found in which they will vanish.

$$\hat{K}(\mathbf{q}, \mathbf{z}) = -\frac{\hbar^2}{2} \left(\sum_{k,l=1}^{3H-6} g_{kl} \frac{\partial^2}{\partial q_k \partial q_l} + \sum_{k=1}^{3H-6} h_k \frac{\partial}{\partial q_k} \right)$$

$$+ \frac{\hbar^2}{2} \left(\sum_{\alpha\beta} \kappa_{\alpha\beta} \hat{l}_\alpha \hat{l}_\beta + \sum_\alpha \lambda_\alpha \hat{l}_\alpha \right) \tag{18}$$

in which $\mathbf{g}$ is a $3H-6$ by $3H-6$ matrix given by

$$\mathbf{g} = \sum_{i=1}^{H} m_i^{-1} {Q^i}^T Q^i \tag{19}$$

In the terms linear in the derivatives of the coordinates

$$h_k = \sum_{i=1}^{H} m_i^{-1} \left(\sum_\beta ({\Omega^i}^T M^\beta Q^i)_{\beta k} + \sum_{l=1}^{3H-6} \left({Q^i}^T \frac{\partial}{\partial q_l} Q^i \right)_{lk} \right) \tag{20}$$

The potential energy operator is

$$V(\mathbf{q}, \mathbf{z}) = \frac{e^2}{8\pi\varepsilon_0} \sum_{i,j=1}^{L} {}' \frac{1}{|\mathbf{z}_j - \mathbf{z}_i|} + \frac{e^2}{8\pi\varepsilon_0} \sum_{i,j=1}^{H} {}' \frac{Z_i Z_j}{r_{ij}} - \frac{e^2}{4\pi\varepsilon_0} \sum_{i=1}^{H} \sum_{j=1}^{L} \frac{Z_i}{r'_{ij}}$$

or

$$V(\mathbf{q}, \mathbf{z}) = V^e(\mathbf{z}) + V^n(\mathbf{q}) - V^{ne}(\mathbf{q}, \mathbf{z}) \tag{21}$$

Here r'_{ij} is an electron-nucleus distance and r_{ij} is an internuclear distance.

It is clear from what has gone before and indeed from general group theoretical arguments, that the eigenfunctions of the internal motion problem, ignoring the overall translation, can be written in the form

$$\Psi^{J,M}(\boldsymbol\phi, \mathbf{q}, \mathbf{z}) = \sum_{k=-J}^{+J} \Phi_k^J(\mathbf{q}, \mathbf{z}) |JMk\rangle \tag{22}$$

where the internal coordinate function on the right side cannot depend on M because, in the absence of a field, the energy of the system does not depend on M.

One can eliminate angular motion from the problem and write an effective Hamiltonian within any (J, M, k) rotational manifold, that depends only on the internal coordinates. The transformation from the $\mathbf{x}_i$ to the eulerian angles and the internal coordinates is non-linear and hence has a jacobian which is a function of coordinates. The non-linearity is a topological consequence of any transformation that allows rotational motion to be separated (see Schutz[4]) and there is always some conformation of the particles that causes the jacobian to vanish. Clearly where the jacobian vanishes, the transformation is undefined. This failure manifests itself in the Hamiltonian by the presence of terms which diverge unless, acting on the wavefunction, they vanish. This can occur either by cancellation or by the wavefunction itself being vanishingly small in the divergent region. The origin of these divergences is not physical: they arise simply as a consequence of the choice of coordinates. The important point is that the non-linear transformation cannot be globally valid. As it has only local validity, one can at most derive a local Hamiltonian that is valid within a particular domain.

A permutation of the variables describing a set of identical particles will naturally leave the centre of mass coordinate unchanged but it will also induce changes in the orientation variables and in the internal coordinates. It can be shown that a permutation of identical particles can produce from a given set of orientation variables, a new set of orientation variables which may be functions of the original orientation variables and the internal coordinates, However such a permutation can change an internal coordinate only into a function of internal coordinates. It is possible in principle to choose a set of orientational variables that are invariant under permutations of identical particles, but the commonly made choices are not invariant. Similarly the commonly made choices of internal coordinates are not invariant. Unless choice of invariant orientation variables is made, however, the form of the internal motion Hamiltonian within any rotational manifold, cannot be invariant either. In this case the idea of vibration-rotation separation becomes a problematic one.

Let us consider these problems in the light of putting a chemical molecule into the quantum mechanics.

Historically in quantum mechanics, the nuclei have always been treated on a different footing from the electrons. This difference is a hangover from the old quantum theory but it is usually justified now by appeal to the pioneering work of Born and Oppenheimer[5]. The argument of that paper, put informally, is that because the nuclei are so much heavier than the electrons, they must move much more slowly. So, to a first approximation it is sensible to hold (or clamp) them at a sequence of fixed positions and to solve the problem for the fast-moving electrons as a function of the nuclear positions. The nuclear motions can then be accommodated as perturbations. In fact the perturbation argument of the original paper is now seldom quoted but rather the later work of Born which is most readily accessible as Appendix VIII in Born and Huang[6]. (See also Slater[7].)

Invoking the variable split made earlier. the electronic wave function is then provided by a solution of the clamped nuclei electronic Hamiltonian

$$\hat{H}^{en}(\mathbf{a}, \mathbf{x}^e) = -\frac{\hbar^2}{2m} \sum_{i=1}^{L} \nabla^2(\mathbf{x}_i^e) - \frac{e^2}{4\pi\varepsilon_0} \sum_{i=1}^{H} \sum_{j=1}^{L} \frac{Z_i}{|\mathbf{x}_j^e - \mathbf{a}_i|} + \frac{e^2}{8\pi\varepsilon_0} \sum_{i,j=1}^{N}{}' \frac{1}{|\mathbf{x}_i^e - \mathbf{x}_j^e|} \tag{23}$$

This Hamiltonian is obtained from the original one (1) by assigning the values $\mathbf{a}_i$ to the nuclear variables $\mathbf{x}_i^n$, hence the designation *clamped nuclei* for this form. Within the electronic problem each nuclear position $\mathbf{a}_i$ is treated as a parameter. It has eigenfunctions of the form $\phi(\mathbf{a}, \mathbf{x}^e)$ and energies of the form $E(\mathbf{a})$. Thus both the eigenfunctions and the eigenvalues contain the nuclear variables as parameters. The energy here is commonly called the electronic energy and the eigenfunction, the electronic wave function. For solution of the entire problem, the electronic wave function must be available for all values of the parameters.

If it were the case that the solution to the full problem could be written as a product of the solution to the clamped nuclei problem and a nuclear motion function

$$\psi(\mathbf{x}^n, \mathbf{x}^e) = \Phi(\mathbf{x}^n)\phi(\mathbf{x}^n, \mathbf{x}^e)$$

the function, $\Phi(\mathbf{x}^n)$ could be determined as a solution of an effective nuclear motion equation, obtained from the full equation (3) by multiplying from the left by $\phi^*(\mathbf{x}^n, \mathbf{x}^e)$ and integrating over all $\mathbf{x}_i^e$. Ignoring the effects of the nuclear derivative operators on the electronic wave function, the effective nuclear motion problem so constructed, has a potential that is just the sum of the electronic energy and the nuclear repulsion energy. This potential is usually referred to as the potential energy surface.

It is clear that there is something a little fishy about this approach, for the electronic energy is invariant under all uniform translations and all rigid rotation-reflections of the geometrical figure specified by the particular choice of **a**. So It cannot be a function of all the nuclear coordinates. In fact it can be shown that it is a function of only $3H - 6$ internal coordinates formed from the nuclear variables (excluding the case where $H = 2$). There are unfortunately, formidable technical problems in the way of making a precise argument here and connecting the solutions of the clamped nuclei electronic Hamiltonian in an unambiguous fashion with solutions to the full problem.

Even at a qualitative level it is clear enough that any argument made here must depend on the magnitude of the electronic energy and the details of its behaviour as a function of geometry and the magnitude of the separation between electronic energy levels. To some extent however it is possible to justify the simple product approach for an appropriately small collection of nuclei at suitably low energies near sufficiently deep minima on the potential surface, if this surface is sufficiently well separated from any other potential surface. Assuming it justified, then a molecule is identified as occurring at where there is a local minimum on the potential energy surface. This minimum is then thought of as specifying an equilibrium geometry for a molecule.

This apparent simplification of the problem is not, however, without its price. In the clamped nuclei Hamiltonian (23), the nuclei are identified in order to define the coordinate frame for the electronic calculation. A typical "point" on a potential energy surface occurs at an identifiable nuclear geometry. But if some of the nuclei are identical then geometries resulting from permutations of such nuclei require equal consideration. In practice, as first noted in print by Berry,[8] they do not receive it. An attempt to justify this was made by Longuet-Higgins.[9] He postulated that any permutation that could be realised only by dis-assembling the geometrical figure of a molecule, could be achieved only by surmounting exceptionally high barriers on the potential energy surface and would, therefore be unimportant. Such a permutation was therefore called *un-feasible*. Only permutations that could realised by orthogonal transformations of the rigid geometrical figure were important or *feasible*. This approach has been extremely influential and detailed expositions of it can be found in Ezra[10] and Bunker.[11] Although this way of looking at things does allow us to understand why point group symmetry is such an effective means of classification in molecular spectroscopy, it still leaves some puzzles. Suppose that the potential surface argument is appropriate. In the absence of a calculation, whether there is a barrier or not in realising a permutation, is largely a matter of opinion. It may well be one on which there can be widespread agreement but that may simply be because those agreeing, recognise the need to achieve a particular group symmetry to explain observations. It is not clear either that the potential argument is appropriate. The idea of un-feasibility rests on the notion that the permutation is a real motion of particles in the potential computed from a clamped nuclei calculation. But the idea that a permutation is a real motion of particles is an incongruous one, from a mathematical point of view, as is the idea of un-feasibility, outside the single product approximation for the wave function and hence a single uncoupled potential function.

It remains unclear how best to deal with these difficulties. They add to the problems arising from permutations which are adumbrated above. Neither are they the only problems. It is not clear whether this separation is the best way to consider systems consisting of large numbers of nuclei and electrons, neither is it clear that it is the best way of dealing with systems at high energies or at small separations between or even crossings of, potential energy surfaces.

In other words, although number of particles, together with their masses and charges are intrinsic properties of the system, it is not really possible to use them to identify the molecule *a priori*. It is necessary to use extrinsic criteria. We shall begin our discussion of such criteria by considering the isolated molecule.

3. THE ISOLATED MOLECULE

If it is assumed that the problem specified by

$$\hat{K}(\mathbf{z}) + V^e(\mathbf{z}) - V^{ne}(\mathbf{q}, \mathbf{z})$$

can be mapped into the clamped nuclei electronic problem and if that problem has been solved to yield a potential for nuclear motion, but that otherwise the electrons are not involved in the nuclear motion, then we can concentrate on the nuclear motion problem. In doing this we confine ourselves to accounting for what is usually considered to be just one aspect of general molecular behaviour. This is the response to radiation in the infrared and longer wavelength regions. We could extend the picture that we present below to deal, in a rather *ad hoc* way, with nuclear motion as a perturbation effect on the response to low intensity radiation at visible and shorter wavelengths. But it is still quite unclear how to or indeed, whether. it is possible to modify the picture presented below to deal with the sort of high intensity radiation that gives rise to photo-electron spectroscopy. It must be borne in mind that we tailor our Hamiltonian to deal with a particular kind of experimental results. What we do can be supported or not, only in terms of the success achieved.

If we do neglect the effect of electronic motion on the nuclear motion then (8) simplifies to give

$$\hat{K}(\boldsymbol{\phi}, \mathbf{q}) + \hat{K}(\mathbf{q}) \tag{24}$$

where (10) simplifies to

$$\hat{K}(\boldsymbol{\phi}, \mathbf{q}) = \frac{1}{2}\left(\sum_{\alpha\beta} \kappa_{\alpha\beta} \hat{L}_\alpha \hat{L}_\beta + \hbar \sum_\alpha \lambda_\alpha \hat{L}_\alpha\right) \tag{25}$$

and (18) simplifies to

$$\hat{K}(\mathbf{q}) = -\frac{\hbar^2}{2}\left(\sum_{k,l=1}^{3H-6} g_{kl} \frac{\partial^2}{\partial q_k \partial q_l} + \sum_{k=1}^{3H-6} h_k \frac{\partial}{\partial q_k}\right) \tag{26}$$

Using the form (22) one can eliminate angular motion from the problem and write an effective Hamiltonian within any (J, M, k) rotational manifold, that depends only on the internal coordinates. The matrix elements with respect to the angular variables of the operators that depend only on the q_k are trivial. Thus

$$\langle J'M'k' \mid \hat{K}(\mathbf{q}) + V \mid JMk\rangle = \delta_{J'J}\delta_{M'M}\delta_{k'k}(\hat{K}_I + V) \tag{27}$$

in which V is the potential which arises as the sum of the electronic and nuclear repulsion energy. In what follows explicit allowance for the diagonal requirement on J and M will be assumed and the indices suppressed to save writing. Similarly the fact that the integration implied is over $\boldsymbol{\phi}$ only will be left implicit.

To treat the operator containing the angular terms is much more complicated and best done by re-expressing the components of $\hat{\mathbf{L}}$ in terms of $\hat{L}_\pm(\boldsymbol{\phi})$ and $\hat{L}_z(\boldsymbol{\phi})$. When this is done

$$\langle JMk' \mid \hat{K}(\phi, \mathbf{q}) \mid JMk \rangle =$$

$$\frac{\hbar^2}{4}(b_{+2}C^+_{Jk+1}C^+_{Jk}\delta_{k'k+2} + b_{-2}C^-_{Jk-1}C^-_{Jk}\delta_{k'k-2})$$

$$+\frac{\hbar^2}{4}(C^+_{Jk}(b_{+1}(2k+1)+\lambda_+)\delta_{k'k+1} + C^-_{Jk}(b_{-1}(2k-1)+\lambda_-)\delta_{k'k-1})$$

$$+\frac{\hbar^2}{2}((J(J+1)-k^2)b + b_0k^2 + \lambda_0 k)\delta_{k'k} \qquad (28)$$

In this expression

$$C^{\pm}_{Jj} = [J(J+1)-j(j\pm1)]^{\frac{1}{2}}$$

$$b_{\pm2} = (\kappa_{xx}-\kappa_{yy})/2 \pm \kappa_{xy}/i$$

$$b_{\pm1} = \kappa_{xz} \pm \kappa_{yz}/i$$

$$b = (\kappa_{xx}+\kappa_{yy})/2 \qquad\qquad b_0 = \kappa_{zz} \qquad (29)$$

and in terms of the λ_α in (15) λ_0 is λ_z and the $\lambda_\pm$ are

$$\lambda_\pm = (\lambda_x \pm \lambda_y/i) \qquad (30)$$

The Spectroscopic Molecule

When thinking of a system containing at most a dozen or so nuclei it is usual to realise this nuclear motion Hamiltonian in terms of the approach begun by Eckart[12] and brought to full fruition in the work of Watson.[13] The resulting quantum mechanical form is particularly appropriate for interpreting molecular spectra and so will be thought of as defining the spectroscopic molecule. The physical picture is one in which there is an isolated deep minimum on the potential surface at a particular nuclear geometry, the *equilibrium* geometry or structure. To describe this geometry Eckart begins from the H redundant nuclear coordinates

$$\mathbf{x}^n_i - \mathbf{X} = \mathbf{Cr}_i$$

where $\mathbf{X}$ is the centre of nuclear mass coordinate. Thus

$$\sum_{i=1}^{H} m_i \mathbf{r}_i = 0$$

and a matrix $\mathbf{C}$ is chosen as in (4) to define a set of cartesians in the frame fixed in the body such that the equilibrium structure is specified by

$$\mathbf{r}_i = \mathbf{a}_i$$

where the $\mathbf{a}_i$ are constant matrices. It as at this juncture that the nuclei are identified and, as noted above, that the full permutational symmetry of the problem is broken. The resulting Hamiltonian has only the symmetry of the point group of the equilibrium geometry.

The $\mathbf{r}_i$ are completely expressible in terms of a set of $3H-6$ internal coordinates. The internal coordinates are expressed in terms of displacements from the equilibrium geometry by

$$q_k = \sum_{i=1}^{H}\sum_{\alpha} b_{k\alpha i}(\mathbf{r}_i - \mathbf{a}_i)_\alpha, \ k = 1,2,\cdots 3H-6$$

where the elements $b_{k\alpha i}$ are simply constants which may be regarded as components of a column matrix $\mathbf{b}_{ki}$. The range of these coordinates is $(-\infty, \infty)$. The precise conditions specifying $\mathbf{C}$ are

$$\sum_{i=1}^{H} m_i(\hat{\mathbf{a}}_i(\mathbf{r}_i - \mathbf{a}_i)) \equiv \sum_{i=1}^{H} m_i \vec{\mathbf{a}}_i \times (\vec{\mathbf{r}}_i - \vec{\mathbf{a}}_i) = 0$$

provided that the $\mathbf{a}_i$ do not define a line. These constraint conditions are on the components of the mass weighted sum over all the vectors, of the vector products of the equilibrium geometry vectors with the displacement vectors. In classical mechanics the vanishing of these components would be interpreted as the system having no internal angular momentum at the equilibrium geometry.

If the displacements are sufficiently small then the potential for nuclear motion can be expanded about the equilibrium geometry as a power series in the q_k

$$V(\mathbf{q}) = V(\mathbf{a}) + \frac{1}{2} \sum_{kl=1}^{3H-6} F_{kl} q_k q_l + \cdots$$

The first term in this expression is the sum of the electronic and the nuclear repulsion energy at the equilibrium geometry. The linear term is absent because the first derivatives of the potential vanish at the equilibrium geometry, by definition. The matrix $\mathbf{F}$ is composed of the second derivatives of the potential evaluated at the equilibrium geometry. The matrix of second derivatives is, in the context of power series often called the Hessian matrix and so $\mathbf{F}$ is often called the Hessian at the minimum. It is a symmetric positive definite matrix and its elements are usually called the *force constants* for the problem. This usage arises because they have units of $[\text{force}][\text{distance}]^{-1}$ and in the classical texts on spectroscopy are always quoted in millidynes per Ångström.

Internal coordinates must be linearly independent and so the $\mathbf{b}_{ki}$ must be linearly independent and in order for an inverse transformation to exist between the internal coordinates and the coordinates in the frame fixed in the laboratory, it is required that

$$\sum_{i=1}^{H} \mathbf{b}_{ki} = 0 \qquad \sum_{i=1}^{H} \hat{\mathbf{a}}_i \mathbf{b}_{ki} \equiv \sum_{i=1}^{H} \vec{\mathbf{a}}_i \times \vec{\mathbf{b}}_{ki} = 0$$

The Hamiltonian derived as outlined above using these general internal coordinates is really very cumbersome and extremely difficult to use, but Watson[13] showed (see also Louck[14] for an approach rather more like ours) that it could be simplified by a particular choice of orthogonal internal coordinates, and by incorporating the internal coordinate part of the jacobian into the operator. This last trick is analogous to the fairly familiar process when working in spherical polars, for example, where the radial volume element $r^2 dr$ can be reduced to dr by writing the trial wavefunction $\psi(r)$ as $r^{-1}P(r)$ and modifying the Hamiltonian to refer to $P(r)$. This modification changes the derivative terms in the operator by $\partial/\partial r \rightarrow (\partial/\partial r - 1/r)$ and so on but alters none of the multiplicative or $\partial/\partial\theta$ terms. The resulting Hamiltonian is often said to be in *manifestly hermitian* form. Particular examples of this kind of construction can be found in Watson[13] and in Louck[14] while a general account is given in Section 35 of Kemble[15].

With Watson's choice the $\mathbf{b}_{ki}$ are written $\sqrt{m_i}\mathbf{l}_{ki}$ and it is required that

$$\sum_{i=1}^{H} \mathbf{l}_{ki}^T \mathbf{l}_{mi} = \delta_{km}, \quad \sum_{i=1}^{H} \sqrt{m_i}\mathbf{l}_{ki} = 0, \quad \sum_{i=1}^{H} \sqrt{m_i}\hat{\mathbf{a}}_i\mathbf{l}_{ki} = 0$$

The internal coordinates so defined are usually written as Q_k where, explicitly

$$Q_k = \sum_{i=1}^{H} \sqrt{m_i}\mathbf{l}_{ki}^T(\mathbf{r}_i - \mathbf{a}_i)$$

Coordinates satisfying the above conditions can always be chosen to diagonalise a quadratic approximation to the potential as well. If they are so chosen, then they are called *normal* coordinates and are the basis of the simplest first-order approach to describing the vibration-rotation motion of the molecule.

The kinetic energy operators for the spectroscopic molecule can be written as

$$\hat{K}(\phi, \mathbf{Q}) = \frac{1}{2} \sum_{\alpha\beta} \mu_{\alpha\beta} (\hat{L}_\alpha + \hat{\pi}_\alpha)(\hat{L}_\beta + \hat{\pi}_\beta)$$

$$\hat{K}(\mathbf{Q}) = -\frac{\hbar^2}{2} \sum_{k=1}^{3H-6} \frac{\partial^2}{\partial Q_k^2} - \frac{\hbar^2}{8} \sum_\alpha \mu_{\alpha\alpha} \tag{31}$$

in which μ is a special form of the inverse generalised inertia tensor κ in (10) defined as

$$\mu = \mathbf{I''}^{-1} \mathbf{I}^0 \mathbf{I''}^{-1}$$

Here $\mathbf{I}^0$ is the inertia tensor for the molecule at the equilibrium geometry.

$$\mathbf{I}^0 = \sum_{i=1}^{H} m_i \hat{\mathbf{a}}_i^T \hat{\mathbf{a}}_i$$

and so is a constant matrix, while

$$\mathbf{I''} = \sum_{i=1}^{H} m_i \hat{\mathbf{r}}_i^T \hat{\mathbf{a}}_i = -\sum_{i=1}^{H} m_i \hat{\mathbf{r}}_i \hat{\mathbf{a}}_i$$

and is a matrix which allows for the instantaneous geometry changes. The operator $\hat{\pi}_\alpha$ is defined as

$$\hat{\pi}_\alpha = \frac{\hbar}{i} \sum_{k,l=1}^{3H-6} \zeta_{kl}^\alpha Q_k \frac{\partial}{\partial Q_l}$$

in which

$$\zeta_{kl}^\alpha = \sum_{i=1}^{H} (\hat{\mathbf{l}}_{ki} \mathbf{l}_{li})_\alpha$$

This operator is often called a component of internal angular momentum but this is rather a misnomer, for it has none of the properties of angular momentum. It is actually a component of the Coriolis coupling operator and thus the constants ζ_{kl}^α are quite properly given the usual designation, Coriolis coupling constants.

In the original presentation by Watson, the components of the total angular momentum operators are denoted $\hat{\Pi}_\alpha$ and obey the anomalous commutation conditions. Here we have followed Louck and used angular momentum operators that obey the standard commutation conditions. The Watson form is achieved from the present one by making the substitution

$$(\hat{L}_\alpha + \hat{\pi}_\alpha) \to -(\hat{\Pi}_\alpha - \hat{\pi}_\alpha)$$

If the displacements are small, then the Q_k are vanishingly small and so the Coriolis coupling operator vanishes and the inverse generalised inertia tensor becomes μ^0, the inverse of the equilibrium inertia tensor. In these circumstances one can choose the frame fixed in the body so that the equilibrium inertia tensor is diagonal (the principal axis choice). Also an expansion of the potential up to quadratic terms in the coordinates, will be adequate. If this is the case then the equations for the kinetic energy above simplify to yield a Hamiltonian

$$\hat{H}(\phi, \mathbf{Q}) = \hat{K}(\phi) + \hat{H}(\mathbf{Q})$$

in which

$$\hat{K}(\phi) = \frac{1}{2}\sum_{\alpha} \mu_{\alpha\alpha}^{0} \hat{L}_{\alpha}^{2} \qquad (32)$$

$$\hat{H}(\mathbf{Q}) = -\frac{\hbar^2}{2}\sum_{k=1}^{3H-6} \frac{\partial^2}{\partial Q_k^2} + \frac{1}{2}\sum_{k=1}^{3H-6} \lambda_k Q_k^2 \qquad (33)$$

where the zero of the potential energy in (33) has been chosen to incorporate the constant terms. It has also been assumed that the coordinates are normal ones, chosen to diagonalise the quadratic approximation to the potential. Thus the λ_i are the eigenvalues of the quadratic form F. All the eigenvalues must be positive for a stable molecule because the matrix of second derivatives of the internal coordinates evaluated at a minimum (the Hessian at the minimum) must be positive definite. If for any reason the expansion is made not about the equilibrium geometry, but rather at some other geometry, then it is possible to get negative values of λ_i and hence imaginary vibration frequencies.

Because the elements of $\boldsymbol{\mu}^0$ are constants then (32) is just the Hamiltonian for an asymmetric top. The rotational matrix element in (28) simplifies to

$$\langle JMk' \mid \hat{K}(\phi) \mid JMk \rangle =$$

$$\frac{\hbar^2}{4}\left(\frac{(\mu_{xx}^0 - \mu_{yy}^0)}{2}\left(C_{Jk+1}^{+}C_{Jk}^{+}\delta_{k'k+2} + C_{Jk-1}^{-}C_{Jk}^{-}\delta_{k'k-2}\right)\right)$$

$$+\frac{\hbar^2}{2}\left(\frac{(\mu_{xx}^0 + \mu_{yy}^0)}{2}\left(J(J+1) - k^2\right) + \mu_{zz}^0 k^2\right)\delta_{k'k} \qquad (34)$$

The $2(J+1)$ dimensional secular problem composed of these matrix elements cannot generally be solved to give an energy expression in closed form but the rotational wave functions solutions are of the form

$$^{M}\chi_{\tau}^{J}(\phi) = \sum_{k=-J}^{k=J} c_{\tau k}^{J}|JMk\rangle, \ \tau = -J, -J+1, \cdots, J$$

The $c_{\tau k}^{J}$ are constant coefficients and each rotational wave function is associated with an energy $E_{J\tau}$.

If two of the equilibrium moments of inertia are the same (the symmetric top case) then these may be designated as x and y and the first term in (34) vanishes. The energy is then given by the last term in (34) and the $|JMk\rangle$ are individually angular eigenfunctions. Thus for the symmetric top, k is a good quantum number. These matters are treated in standard texts on molecular spectroscopy.

The Hamiltonian (33) simply represents a sum of non-interacting Harmonic oscillators, each with a wavefunction of the form

$$\psi_{n_i}(Q_i) = N_i e^{-\frac{\alpha_i Q_i^2}{2}} H_{n_i}(\sqrt{\alpha_i}Q_i)$$

where

$$\alpha_i = \frac{\sqrt{\lambda_i}}{\hbar} \equiv \frac{\omega}{\hbar}$$

37

and the energy of the oscillator is

$$\varepsilon_{n_i} = \left(n_i + \frac{1}{2}\right)\hbar\omega \equiv \left(n_i + \frac{1}{2}\right)h\nu$$

The full vibrational wave function is then

$$\Psi(\mathbf{Q}) = \prod_{i=1}^{3H-6} \psi_{n_i}(Q_i)$$

and the total vibrational energy of the system is just

$$E_v = \sum_{i=1}^{3H-6} \varepsilon_{n_i}$$

The wavefunction for the nuclear motion part of the problem arises from the simplification of (22) and is a product

$$^M\chi_\tau^J(\phi)\Psi(\mathbf{Q})$$

and the wavefunction for the full problem in the single product approximation is

$$\psi(\mathbf{x}^n, \mathbf{x}^e) \Rightarrow T(\mathbf{X}_T)\,^M\chi_\tau^J(\phi)\Psi(\mathbf{Q})\phi(\mathbf{a},\mathbf{z})$$

in which $\phi(\mathbf{a},\mathbf{z})$ is the electronic wavefunction taken at the equilibrium nuclear geometry. The total energy of the molecule in this approximation is

$$E = E_T + E_{J\tau} + E_v + V(a)$$

The translational energy E_T is usually ignored as is the translational wavefunction and the fact that in this approximation the energy is the sum of an electronic and a rotational and a vibrational part is often (quite wrongly) said to specify the Born Oppenheimer approximation.

So to compute the properties of isolated spectroscopic molecule in quantum mechanics, the following separate problems need to be solved. We must first perform an electronic structure calculation and identify the equilibrium geometry. At that geometry we must calculate, in a convenient coordinate system, the Hessian for the nuclear variables. The electronic structure calculation must therefore be good enough to distinguish the lowest energy minimum from any others that might occur locally and at the lowest energy minimum, be good enough to yield a decent equilibrium geometry and a potential surface that is parallel to the exact one for small displacements about this geometry. For a closed-shell system an LCAO-MO-SCF calculation is generally believed to be adequate if a suitably large basis is chosen. If it is necessary to go further, as it sometimes is to make sure that the true minimum has been found, perturbation techniques such as MP2 are used.

The Hessian determined from the electronic structure calculation can be diagonalised to yield the eigenvalues λ_i and hence the harmonic the vibrational frequencies of the molecule. The eigenvectors express the normal coordinates in terms of the coordinates used to evaluate the Hessian. The equilibrium inertia tensor can be calculated from the nuclear coordinates that specify the equilibrium geometry. This tensor can then be diagonalised to give the moments of inertia and the secular problem with matrix elements given by (34) can be solved to yield the rotational energy levels. The Hessian is usually constructed using the cartesian coordinates in which the clamped nuclei calculations are performed. In this case then, the Hessian is of dimension $3H$ but six of its eigenvalues are zero corresponding to the three possible uniform translations and the three possible rigid rotations of the equilibrium geometry.

Provided that this simple picture allows a reasonable interpretation of the molecular spectrum then it is worthwhile refining it by passing beyond the harmonic approximation and allowing for a more realistic representation of the potential. Also by allowing for some non-rigidity in the rotor. All the corrections at this level can be treated by first and second order perturbation theory. At this level then, the isolated molecule problem can clearly be solved. However, if vibration-rotation interaction is important then perturbation theoretic treatments are often inadequate. Also if the electronic motion interacts with the nuclear motion, a state of affairs usually called the breakdown of the Born Oppenheimer approximation, it is less clear how best to deal with it. Indeed in these circumstances the isolated spectroscopic molecule problem must be considered still unsolved.

However it should be remembered that the picture given above pre-dated any possibility of decent calculations. Following Eckart's paper [12] the scheme outlined above was developed and used to interpret spectra, that is, it was assumed to be valid and experiment was used to parametrise it. In this process it was discovered that spectra could be interpreted in terms of molecular properties that were, in good approximation, transferable. Thus the rotational spectrum of a pair of molecules that would be thought of as having chemically similar bonds can be interpreted by means of an equilibrium geometry in which the common bonds are the same length. Similarly the vibrational spectrum of both can be interpreted by constructing a potential quadratic in terms of bond lengths and bond angles as internal coordinates with characteristic force constants for each quadratic term. For example, it is found that $C-H$ stretching force constant (one quadratic in the displacement along the bond) can be given roughly the same value in whatever molecule it occurs. This picture is the one that experimentalists use and it is in terms of this picture that a theoretician is expected to present results. This does not necessarily mean that calculations must be done within this scheme, but it does mean that results are best presented in terms of it to make contact with the standard picture.

Now if the environment is conceived of as consisting of other isolated spectroscopic molecules moving essentially independently of each other (modelling, perhaps, a dilute gas at low pressure) then the standard methods of statistical mechanics are sufficient to produce pretty decent values for heat capacities and the like from molecular properties calculated or assigned according to the Eckart model. Beyond this, a simple picture of discreet molecules each described in the Eckart manner but interacting through, say, a Lennard-Jones potential, often gives very satisfactory results for bulk properties like viscosity and compressibility. However if the environment is regarded as other molecules forming a liquid so that the whole system is a solution, then the difficulties are much greater. And they are even greater if the system is a solid. It is not yet clear whether it would be at all useful to work towards an account of such systems through the Eckart Hamiltonian for an isolated molecule. It seems likely then that the approach outlined in this section is probably not going to be helpful when considering really large molecules. Once the usual state of the system is a liquid or a solid without a significant gas phase component, then looking at its wave function in the manner outlined above, is possibly waste of time. It seems very unlikely, for example, that it would be helpful to consider even a molecular solid in terms of individual intermolecular interactions but rather in terms of an extended potential whose origin can formally be attributed to such intercations.

It still seems quite useful to assume that the internal motions can be separated off and to consider an aspect of the behaviour of such molecules as arising from their tendency to achieve a minimum energy geometry on a potential energy surface with many minima in it. But, in line with the comments made at the beginning of the section, if one wants to describe a molecule as part of a system or in an environment, one simply has to build a Hamiltonian that seems likely to do the job and in doing so, it is not clear that one will be able to hang on to all the aspects of a molecule that have emerged from a Hamiltonian designed to do another job.

4. THE ELECTROMAGNETIC FIELD AS AN ENVIRONMENT

The problem to be solved here is how to treat fields in quantum mechanics in order to describe the results of the many interesting experiments that result from molecules responding to applied fields. Among the effects that result from the presence of fields are the Stark effect, the Zeeman effect, the Kerr effect and the Cotton-Mouton and the Faraday effects. Intuitively we tend to believe that fields usually have only a small effect on molecules. So we tend to believe that in the esr or nmr experiment we are essentially looking at signals resulting from the molecule, perturbed only to a small extent.

The System in an Electromagnetic Field

In quantum mechanics the electromagnetic operators (see McWeeny[16], Moss[17]) are multiplicative and are the scalar potential ϕ and the vector potential A associated with the electric field strength E and the magnetic flux density B according to

$$E = -\vec{\nabla}\phi - \frac{\partial A}{\partial t}$$

$$B = \vec{\nabla} \times A$$

The electromagnetic operators in the Hamiltonian represent interactions between the field and the particles and so are written as $\phi(x_i)$ and $A(x_i)$ and associated with the i–th particle. In the presence of a field, it can be shown that the Hamiltonian (1) becomes

$$\hat{H}(x,\phi,A) = \sum_{i=1}^{N} \frac{1}{2m_i}(\frac{\hbar}{i}\vec{\nabla}(x_i) - eZ_i A(x_i)) \cdot (\frac{\hbar}{i}\vec{\nabla}(x_i) - eZ_i A(x_i))$$

$$+ \sum_{i=1}^{N} eZ_i\phi(x_i) + \frac{e^2}{8\pi\varepsilon_o} \sum_{i,j=1}^{N} {}^{'}\frac{Z_i Z_j}{x_{ij}} \qquad (35)$$

Expanding the first term gives

$$\hat{H}(x,\phi,A) = \hat{H}(x) + \hat{H}'(x,\phi,A)$$

where $\hat{H}(x)$ is given by (1) and

$$\hat{H}'(x,\phi,A) =$$

$$\sum_{i=1}^{N} \frac{1}{2m_i}(-eZ_i\frac{\hbar}{i}(\vec{\nabla}(x_i) \cdot A(x_i) + A(x_i) \cdot \vec{\nabla}(x_i)) + e^2 Z_i^2 |A(x_i)|^2) + \sum_{i=1}^{N} eZ_i\phi(x_i) \qquad (36)$$

It is usual to write

$$\pi(x_i) = (\frac{\hbar}{i}\vec{\nabla}(x_i) - eZ_i A(x_i))$$

and to call it either the mechanical momentum or the gauge invariant momentum of the particle. The reason for the latter name arises from the observation that replacement of the potentials by the new potentials according to

$$A \to A' = A + \vec{\nabla}f$$

$$\phi \to \phi' = \phi - \frac{\partial f}{\partial t}$$

40

where f is an arbitrary function of position and time, leads to exactly the same fields. These equations define a change of gauge.

In what follows, the potentials will be assumed not to depend explicitly on time, so the time partial derivatives will not be considered further. In this case it is easy to see that π has an invariant expectation value under gauge change provided that the original function undergoes a phase change.

$$\psi \to \psi' = e^{i\lambda}\psi$$

with

$$\lambda = \frac{Ze}{\hbar}f$$

It follows therefore that the physical quantities calculated from the Hamiltonian will be gauge invariant as is required.

The freedom that gauge change allows can be used to recast the Hamiltonian in a way that is particularly useful for some purpose or other. Thus because

$$\vec{\nabla} \cdot A\psi = A \cdot (\vec{\nabla}\psi) + \psi(\vec{\nabla} \cdot A)$$

the second term can be made to vanish if f is chosen to make $\vec{\nabla} \cdot A$ zero. This choice of gauge is often called coulomb gauge but a number of other choices are commonly made too.

In the case of a uniform electric field and of a uniform magnetic flux density it is easily seen that

$$A = \frac{1}{2}B \times \vec{x}$$

and

$$\phi = -E \cdot \vec{x}$$

where E and B are constant vectors. Notice that both these quantities depend upon a particular choice of origin for the position vector $\vec{x}$ and that any constant translation (shift of origin) leaves the derived fields invariant. Notice too that A satisfies the coulomb gauge condition. In this case (36) may be re-written as

$$\sum_{i=1}^{N} \frac{1}{2m_i}\left(-eZ_i B \cdot \frac{\hbar}{i}\vec{x}_i \times \vec{\nabla}(x_i) + e^2 Z_i^2 |\frac{B}{2} \times \vec{x}_i|^2\right) - eE \cdot \sum_{i=1}^{N} Z_i\vec{x}_i \tag{37}$$

It might be prudent here to say a little about units and dimensions. The SI unit of electric field strength is the volt metre^{-1} (V m^{-1}) and the unit of magnetic flux density is the tesla (T). The unit of scalar potential is therefore the volt which in the present context is most helpfully regarded as having units of joule coulomb^{-1} (J C^{-1}). The vector potential has units of tesla metre (T m).

In (37) the term linear in the magnetic field is usually interpreted as a field-dipole interaction with the magnetic dipole arising from the vector product term with a coefficient of the form $e\hbar m_i^{-1}$. If m_i is taken as the electron mass then this coefficient has the approximate value of $1.8546 \ 10^{-23}$ J T^{-1} and if m is taken as the proton mass then it is about 2000 times smaller. Commonly occurring B fields are all less than 1 T. The value for the earth's field is about $5 \ 10^{-5}$ T, that at the centre of a circular coil of 1000 turns each of radius 10^{-1} m and carrying 1.5 A is about 10^{-2} T. However that between the pole pieces of a large electromagnet is about 2 T while the latest in superconducting magnets for nmr machines have a maximum field of about 20 T. In our simple molecular picture we should expect low-lying electronic states to be separated by about 10^{-18} J, low-lying vibrational states by about 10^{-21} J, and low-lying rotational states by about 10^{-23} J.

The electric field term is usually interpreted as an electric dipole interaction and the basic unit is ea_o where $a_o = \dfrac{4\pi\varepsilon_o\hbar^2}{me^2}$. This unit is approximately $8.4778\ 10^{-30}$ C m (about 2.54 Debye) and thus the electric field term yields only small effects for most commonly occurring cases. However it can be shown (Thirring[18]) that an electric field makes a qualitative difference to the problem. In the presence of an electric field, no matter how small, the system becomes metastable and has no discrete spectrum at all. It can also be shown that to treat the electric field as a perturbation to the field-free problem gives perturbation series which are divergent. Nevertheless such series seem to give the correct answer in low order and we shall assume that we can use them in what follows.

The Molecule in an Electromagnetic Field

In line with the earlier discussion we shall move directly to consider the problem from the point of view of a frame fixed in the body of the system with

$$x_i^e - X = Cz_i, \quad x_i^n - X = Cr_i \tag{38}$$

and in which

$$X_T = X + M_T^{-1}mCz, \quad z = \sum_{i=1}^{L} z_i$$

and where it is assumed that the r_i are expressed as functions of the q_k.

Under the change to internal coordinates the term linear in B in (37) becomes

$$-\frac{e\hbar B}{2i} \cdot \sum_{i=1}^{N} \left(\frac{Z_i}{m_i} \vec{x}_i \times \vec{\nabla}(x_i) \right) =$$

$$-\frac{eZ_T\hbar B}{2iM_T} \hat{X}_T \frac{\partial}{\partial X_T} +$$

$$-\frac{e\hbar B'}{2iM_T}(\hat{r} + (1 + mM_T^{-1})\hat{z}) C^T \frac{\partial}{\partial X_T}$$

$$-\frac{e\hbar B}{2i} \hat{X}_T C \Big(\sum_{i=1}^{H} \frac{Z_i}{m_i} \frac{\partial}{\partial n_i} - \Big(\frac{Z_H}{M} + \frac{1}{m}\Big) \sum_{j=1}^{L} \frac{\partial}{\partial z_j} \Big)$$

$$+\frac{e\hbar B'm}{2iM_T} \hat{z} \Big(\sum_{i=1}^{H} \frac{Z_i}{m_i} \frac{\partial}{\partial n_i} - \Big(\frac{Z_H}{M} + \frac{1}{m}\Big) \sum_{j=1}^{L} \frac{\partial}{\partial z_j} \Big)$$

$$+\frac{e\hbar B'}{2iM} \hat{r} \sum_{j=1}^{L} \frac{\partial}{\partial z_j}$$

$$-\frac{e\hbar B'}{2i} \sum_{i=1}^{H} \frac{Z_i}{m_i} \hat{r}_i \frac{\partial}{\partial n_i}$$

$$+\frac{e\hbar B'}{2m} \hat{l} \tag{39}$$

where Z_T is the total charge and Z_H the nuclear charge of the system. and

$$r = \sum_{i=1}^{H} Z_i r_i, \quad B_\alpha' = |C| \sum_{\beta} B_\beta C_{\beta\alpha}$$

On the right in (39) the vector B (B') is to be interpreted as a row matrix of cartesian components. The components of B' are those of the magnetic induction expressed in the axis system defined from the laboratory fixed one by means of C. They can be regarded as components expressed in terms of the axis frame rotating with the molecule, as outlined above.

Some further simplification of (39) may be achieved by defining

$$\Omega^H = \sum_{i=1}^{H} \frac{Z_i}{m_i}\Omega^i, \quad Q^H = \sum_{i=1}^{H} \frac{Z_i}{m_i}Q^i$$

then

$$\sum_{i=1}^{H} \frac{Z_i}{m_i}\frac{\partial}{\partial n_i} = \Omega^H(\frac{i}{\hbar}\hat{L} + i\hat{l}) + Q^H\frac{\partial}{\partial q}$$

and by further defining

$$i^H = \sum_{i=1}^{H} \frac{Z_i}{m_i}\hat{r}_i\Omega^i, \quad j^H = \sum_{i=1}^{H} \frac{Z_i}{m_i}\hat{r}_i Q^i$$

we can write the penultimate term in (39) as

$$-\frac{eB'}{2}i^H(\hat{L} + \hbar\hat{l}) - \frac{e\hbar B'}{2i}j^H\frac{\partial}{\partial q}$$

Rewriting this with together with the last term gives

$$-\frac{eB'}{2}i^H\hat{L} + \frac{e\hbar B'}{2}(1/m - i^H)\hat{l} - \frac{e\hbar B'}{2i}j^H\frac{\partial}{\partial q} \tag{40}$$

If the ratio Z_i/m_i were a constant for all i then j^H would be a null matrix and i^H would be a negative constant multiple of the three by three unit matrix. In these circumstances (40) could be treated as an entirely angular momentum term. If m_i is taken in relative atomic mass units and referred to the atom, then the ratio is pretty close $1/2$ for atoms in the first row of the periodic table, discounting the hydrogen atom. The relative atomic mass unit is about 2000 times the electron mass.

The term quadratic in B is

$$\frac{e^2}{8}\sum_{i=1}^{N}\frac{Z_i^2}{m_i}\left(|B|^2|\vec{x}_i|^2 - |B\cdot\vec{x}_i|^2\right) =$$

$$\frac{e^2}{8}p(|B|^2|X_T|^2 - |BX_T|^2)$$

$$-\frac{e^2}{4}\left(\frac{1}{m} - \frac{mp}{M_T}\right)(|B|^2(X_T^T Cz) - (BX_T)(BCz))$$

$$+\frac{e^2}{4}\left(|B|^2 X_T^T C\sum_{i=1}^{H}\frac{Z_i^2}{m_i}r_i - (BX_T)\sum_{i=1}^{H}\frac{Z_i^2}{m_i}(BCr_i)\right)$$

$$+\frac{e^2}{4}\left(\frac{m^2p}{2M_T^2} - \frac{1}{M_T}\right)((|B|^2|z^2| - (BCz)^2))$$

$$-\frac{e^2m}{4M_T}(|B|^2 z^T\sum_{i=1}^{H}\frac{Z_i^2}{m_i}r_i - (BCz)\sum_{i=1}^{H}\frac{Z_i^2}{m_i}(BCr_i))$$

$$+\frac{e^2}{8}|B|^2\left(\sum_{i=1}^{H}\frac{Z_i^2}{m_i}|r_i|^2+\frac{1}{m}\sum_{i=1}^{L}|z_i|^2\right)$$

$$-\frac{e^2}{8}\left(\sum_{i=1}^{H}\frac{Z_i^2}{m_i}(BCr_i)^2+\frac{1}{m}\sum_{i=1}^{L}(BCz_i)^2\right) \tag{41}$$

where

$$p=\sum_{i=1}^{N}\frac{Z_i^2}{m_i}$$

The term in the electric field becomes

$$-eZ_T EX_T+e\left(1+\frac{mZ_T}{M_T}\right)ECz-eECr \tag{42}$$

Now it would obviously be much neater if it were possible to perform a gauge transformation so as to remove from (39) and (41) the terms corresponding to interactions between the translational motion and the internal motion but this can be done only with a rather particular choice of phase factor f (see for example, Moss and Perry[19], and for a general discussion Johnson et al.[20]) and is not helpful in the present discussion. In any case the interaction between the moving system and the fields is a real physical phenomenon whose effects cannot be transformed away.

If the system is not electrically neutral then the first term on the right in (39) gives rise to cyclotron motion for the centre of mass while the first term in (42) gives rise to a centre of mass Stark effect. If the system is electrically neutral ($Z_T=0$) then the first term on the right in both (39) and (42) vanishes. The centre of mass motion then occurs in the second and third term on the right in (39) and in the first two terms on the right in (41). To estimate the size of the effects due to these terms we would expect to calculate their expectation values between translation functions of the type $T(X_T)$. However such functions, if chosen to be eigenfunctions of the free centre of mass problem, are not square integrable. They are

$$T(X_T)=Ae^{iK^T X_T}$$

where K is a constant column matrix. There is no agreed method for dealing with this problem. One possibility is to treat the centre of mass motion as occurring in a box subject to periodic boundary conditions. This is a widely used trick. It is used, for example, when the quantisation of the radiation field is considered and in solid state theory. If the periodic choice is used then the translational wavefunction becomes

$$T(X_T)=d^{-3/2}e^{iK^T X_T},\quad K=\frac{2\pi}{d}\begin{pmatrix}n_x\\n_y\\n_z\end{pmatrix}=\frac{2\pi}{d}n$$

in which d is the length of the side of a cubic box treated as the repeating unit and the n_α are $0,\pm1,\pm2,\ldots$. The integration of X_T is in the range $(0,d)$.

This wave function is a momentum eigenfunction so

$$\frac{\partial}{\partial X_T}T(X_T)=iK\,T(X_T)$$

and the expected value of the centre of mass vector is

$$\langle T|X_T|T\rangle=\frac{d}{2},\quad d=d\begin{pmatrix}1\\1\\1\end{pmatrix}$$

and the expected value of its square is

$$\langle T||X_T|^2|T\rangle = d^2$$

Assuming that this is a satisfactory wave function it follows that, on integrating over X_T, the expected values of the first three terms on the right in (39) are

$$-\frac{eZ_T\hbar}{4M_T}B\hat{1}n$$

$$-\frac{e\hbar B}{4i}d\hat{1}\,C(\sum_{i=1}^{H}\frac{Z_i}{m_i}\frac{\partial}{\partial n_i}-(\frac{Z_H}{M}+\frac{1}{m})\sum_{j=1}^{L}\frac{\partial}{\partial z_j})$$

$$-\frac{e\hbar B'}{2dM_T}(\hat{r}+(1+mM_T^{-1})\hat{z})C^Tn$$

The interaction terms depend on the length d of the box side and this is a very unsatisfactory outcome since this parameter is entirely at our disposal and may be assigned at will. The larger we choose d to be, the larger the first interaction term and the smaller the second, other things being equal. But the first term is unlikely to be intrinsically small and the second term is, so our difficulties are actually compounded by our field choice. Looking at the similar terms in (41) and in (42) it is clear that they present similar difficulties. At present there seems to be no satisfactory way round this problem.

In exceptionally accurate work on atoms (see for example Baye and Vincke[21] but also see Kravchenko et al.[22]) the energy arising from the collective motion alone is subtracted from the total energy and this appears to yield quantities that are independent of the choices made with respect to the translational function. However the cited work does not approach the problem as has been done here. The traditional approach in molecular physics is simply to ignore these terms for neutral systems and to use rather heroic approximations in the case of charged systems. For the present we must simply flag up this area as one in which there are many unsolved problems which we shall ignore and turn our attention to the internal motion part of the problem.

From what has been said above, the last term in (40) is probably pretty small. The remaining angular momentum terms arising from the internal motion of the particles are usually called paramagnetic terms. They are of the form generally used to account for the non-spin part of the Zeeman splitting. Only one part of the middle term in (40) would remain if the clamped nuclei approximation was made yielding

$$\frac{e\hbar B'}{2m}\hat{1}$$

and this exactly the result given in all the books.

The last four terms in (41) are also just internal motion terms. Each part is positive and so is a diamagnetic term and the first two of these terms might be expected to be intrinsically small because of the reciprocal mass factors. Rewriting the last two as a sum of nuclear and electronic parts gives

$$\frac{e^2}{8}(|B'|^2\sum_{i=1}^{H}\frac{Z_i^2}{m_i}|r_i|^2-\sum_{i=1}^{H}\frac{Z_i^2}{m_i}(B'r_i)^2)$$

$$+\frac{e^2}{8m}(|B'|^2\sum_{i=1}^{L}|z_i|^2-\sum_{i=1}^{L}(B'z_i)^2) \tag{43}$$

If the clamped nuclei approximation was made then only the last term above would survive to give the form usually taken for the diamagnetic term.

The internal motion part of the electric field term is

$$eEC((1+mM_T^{-1}Z_T)z - r) \qquad (44)$$

which is of standard electric dipole form and simplifies in an obvious way for a neutral system or in the clamped nucleus approximation.

It must be stressed again that these results depend intrinsically on the internal coordinate choice. They would not be the same in any other choice.

The Spectroscopic Molecule in an Electromagnetic Field

Making Watson's choice for the internal coordinates

$$\Omega^i = m_i\hat{a}_i I''^{-1}, \quad Q^i_{\alpha k} = m_i^{\frac{1}{2}} l_{k\alpha i}$$

and r_i is realised as

$$r_i = a_i + m_i^{-\frac{1}{2}} \sum_{k=1}^{3H-6} l_{ki}Q_k$$

To be strictly compatible with our earlier discussion of the isolated molecule we should add to the derivative term, a part to allow for the incorporation of the jacobian. This is just a multiplicative term and it will be ignored as it is likely to be small except in regions where the jacobian is close to zero. Details of its calculation (which is rather tricky) can be found in Louck[14].

In these coordinates

$$i^H = \sum_{i=1}^{H} Z_i\hat{r}_i\hat{a}_i I''^{-1}, \quad j^H_{\alpha k} = \sum_{i=1}^{H} Z_i m_i^{-\frac{1}{2}}(\hat{r}_i l_{ki})\alpha$$

and these terms simplify in the required way if Z_i is proportional to m_i. If we make the rigid molecule approximation and simply neglect j^H then I'' becomes I^0 and this may be diagonalised to yield principal axes and then the first term in (40) can be written in a way closely analogous to the semi-classical result quoted in Townes and Schawlow[23]. Similarly replacing r_i by a_i in the first term in (43) gives the semi-classical result for rotational diamagnetism and the same replacement in (44) gives the semi-classical result for the dipole moment of a neutral molecule.

We thus have expressions for the molecule-field interactions in the standard coordinate system.

5. CONCLUSIONS

What I hope that we have seen is that the Schrödinger equation for an isolated molecule has to be quite carefully constructed within ordinary quantum mechanics. However we have seen that such an equation can be constructed and that the molecule can, to some extent, be understood within quantum mechanics in terms of an equilibrium geometry defined by a minimum in a potential energy surface, calculated in principle from the ordinary clamped nuclei electronic Hamiltonian. The full molecular Hamiltonian based on this and as realised in the Eckart–Watson approach, certainly seems to to describe the isolated spectroscopic molecule very well.

Although the description of a collection of molecules has not been dealt with, a context has been provided for any such description. It is hoped that it is now clear why any such description is liable to be tricky and open to disagreement (even more so than in the isolated molecule case) because there is nothing in the Schrödinger Hamiltonian as such, that is responsive to such a division. So extrinsic considerations will be of the essence of such descriptions. Furthermore the exigencies of calculation make it certain that in any account, semi-empirical, even empirical and frankly model methods will have to be used.

When we come to consider placing the molecule in an electromagnetic field then even greater care is necessary than it was in the case of the isolated molecule. But we have seen how it can be done, at least for constant fields and we have indicated what the deep intrinsic problems are. The importance of the gauge choice and hence the coordinate choice is a matter of the first importance here. If we wanted to calculate the electronic contribution to the field-internal motion coupling terms using wave functions obtained from a clamped nuclei calculation, then we *must* choose the origin of any electronic coordinate as the centre of nuclear mass. It is only with such a choice that the clamped nuclei Hamiltonian can be placed properly in the present context and in which the coupling expressions are valid. But this is not as easy as it seems because the standard orbital forms take as their origin the nucleus on which they happen to be centred. This is complication has been studied (see for example Hall[24]) and can be dealt with, but it requires care.

We have not attempted to incorporate the radiation field into our discussion and it would not have been too easy to do it. A recent paper by Woolley[25] and a book by Craig and Thirunamachandran[26] cover much of the relevant ground, albeit without considering the problem of translation and rotation separation. Neither have we attempted to deal with collections of molecules in a field. This is an area in which there are few investigations but one which seems fertile ground for future work. There has also recently been a issue of *International Journal of Quantum Chemistry* devoted entirely to the properties of molecules in strong magnetic fields [27].

REFERENCES

1. B. T. Sutcliffe, *Int. J. Quantum Chem.* 58:645 (1996).
2. W. Gans, A. Amman, and J. Boeyens, eds. "Fundamental Principles of Molecular Modelling," Plenum Press, New York and London (1996).
3. B. T. Sutcliffe, *J. Chem. Soc., Faraday Transactions* 89:2321 (1993).
4. B. Schutz. " Geometrical Methods of Mathematical Physics," Cambridge University Press, Cambridge (1980).
5. M. Born and J. R. Oppenheimer, *Ann.der Phys.* 84:457 (1927).
6. M. Born and K. Huang. "Dynamical Theory of Crystal Lattices," Oxford University Press, Oxford (1955).
7. J. C. Slater, *Proc. Nat. Acad. Sci.* 13:423 (1927).
8. R. S. Berry, *Rev. Mod. Phys.* 32:447 (1960).
9. H. C. Longuet-Higgins, *Molec. Phys.* 6:445 (1963).
10. G. Ezra. " Symmetry Properties of Molecules," Lecture Notes in Chemistry **28**, Springer-Verlag, Berlin (1982).
11. P. R. Bunker. " Molecular Symmetry and Spectroscopy," Academic Press, London (1979).
12. C. Eckart, *Phys. Rev.* 47:552 (1935).
13. J. K. G. Watson, *Mol. Phys.* 15:479 (1968).
14. J. C. Louck, *J. Mol. Spec.* 61:107 (1976).
15. E. C. Kemble. "The Fundamental Principles of Quantum Mechanics," McGraw-Hill, New York (1937).
16. R. McWeeny. "Methods of Molecular Quantum Mechanics," 2nd edn., Academic Press, London (1989).
17. R. E. Moss. "Advanced Molecular Quantum Mechanics," Chapman and Hall, London (1973).
18. W. Thirring. "A Course in Mathematical Physics, 3, Quantum Mechanics of Atoms and Molecules," tr. E. M. Harrell, Springer-Verlag, New York (1981).
19. R. E. Moss and A. J. Perry, *Mol. Phys* 23:957 (1992).
20. B. R. Johnson, J. O. Hirschfelder, and K. H. Yang, *Rev. Mod. Phys* 55:109 (1983).

21. D. Baye and M. Vincke, *J. Phys B: At. Mol. Opt. Phys* 23:246 (1990).

22. Yu. P. Kravchenko, M. A. Lieberman, and B. Johansson, *Phys. Rev A* 54:287 (1996).

23. C. H. Townes and A. L. Schawlow. "Microwave Spectroscopy," Dover, New York (1975).

24. G. G. Hall, *Int. J. Quant. Chem.* 7:15 (1973).

25. R. G. Woolley, *Mol. Phys* 88:291 (1996).

26. D. P. Craig and T. Thirunamachandran. "Molecular Quantum Electrodynamics: an Introduction to Radiation-Molecule Interactions," Academic Press, London (1984).

27. *Int. J. Quantum Chem.*, 64:495-635 (1997).

DYNAMIC ASPECTS OF INTERMOLECULAR INTERACTIONS

J. F. Ogilvie *

Department of Chemistry
University of the Witwatersrand
Private Bag 3, P. O. WITS 2050, South Africa

ABSTRACT

Interactions between atoms are traditionally classified as intramolecular or intermolecular, depending in part on the range of distances between atomic centres. The dynamics of intramolecular interactions in aspects 'molecular vibrations' and intramolecular rearrangement or isomerisation are well characterised, but dynamics of intermolecular interactions are less appreciated by chemists, even though related structural aspects are well established. Here I discuss dynamics of intermolecular interactions for systems of varied size, from a simple case involving only two atoms, through more complicated systems involving familiar organic molecular units, to novel extended inorganic systems. An objective is to relate structural and physical properties to a dynamic nature of intermolecular interactions.

1. INTRODUCTION

To intermolecular interactions there are many aspects. Most commonly considered are perhaps, on a microscopic level, such static characteristics as structure (in its varied manifestations) or, on a macroscopic level, energetic or thermodynamic properties; the latter include, intrinsically, such diverse contributions as kinetic and potential energy in intermolecular vibrational modes in condensed phases, and entropic effects. In some cases dynamic aspects of intermolecular interactions confer on a material specific attributes that relate to particular bulk properties or that can be exploited for distinctive practical applications.

In this essay we consider dynamic aspects of intermolecular interactions within systems of three types, related to interests perhaps primarily to physical, organic and inorganic chemists. The first system is essentially gaseous, whereas the others involve condensed phases, liquid and solid. An objective in the selection of instances of systems is illustration of both a diverse nature of intermolecular interactions in these systems and applications of these and related systems.

---

*permanent address: Centre for Experimental and Constructive Mathematics, Simon Fraser University, Burnaby, BC V5A 1S6 Canada; electronic mail to "ogilvie@cecm.sfu.ca"

2. NOBLE GASES

A simple instance of an intermolecular interaction might be found in a sample, in thermal equilibrium at 300 K, consisting of a noble gas. The deviation of actual behaviour from that of an hypothetical ideal gas is attributed customarily to an existence both of finite spatial extension of an atom (or molecule) and of interatomic forces. These forces are occasionally associated with the name of a physicist van der Waals who proposed in 1873 an empirical equation of state that in some measure took account of such forces; that equation is consistent with forces proportional to internuclear distance to an inverse fourth power.[1] London deduced[2] in 1930 that universal dispersion forces, whereby an electric dipole is induced in one atom by another atom possessing an instantaneous electric dipole, are proportional to internuclear distance to an inverse seventh power; greater powers apply to interactions involving other, multipolar moments. For forces proportional to an inverse seventh power to be the dominant effect at moderate internuclear separation, both such atoms are supposed to lack net intrinsic nuclear angular momentum and to be in an electronic state lacking net electronic orbital or intrinsic angular momentum; in this context an internuclear distance is considered moderate if it is somewhat larger than a sum of effective atomic radii but somewhat smaller than the largest wave length of a resonance electronic transition of a separate atom.[3] Instead of force a convenient quantity with which to work is potential energy $V(R)$, a function that describes how the total energy $E(R)$ of a system consisting of two atomic centres with nuclei fixed in position depends on the internuclear distance R and which is a negative gradient of force between these atomic centres. Within a formulation of the virial equation of state of a real gas, a dependence of second virial coefficient on temperature is readily related to interaction between atoms two at a time, and for the third virial coefficient between atoms three at a time etc.[4] A relation between the second virial coefficient $B(T)$ and potential energy is[4]

$$B(T) = -2\pi N_A \int_0^\infty [e^{(-V(R)/k_B T)} - 1] R^2 dR \qquad (1)$$

About 1925 Lennard-Jones considered these systems of noble gases and represented[5] an interaction between two such atoms as a function of potential energy. On the basis of thermodynamic properties of simple gases such as the second virial coefficient, kinetic properties of gases such as thermal conductivity, diffusivity and viscosity, compressibility of a liquid substance or mixture, and structural properties of a crystalline substance such as lattice parameters and their variation with temperature and pressure, one can generate a potential-energy function with internuclear distance and energy spanning a broad range.[6] When one considers quantum-mechanical (or quantal) ramifications of such a function, one expects that a pair (or couple) of atoms belonging to elements in group 18 of the periodic chart has discrete energies associated with vibrational motion of nuclei in a set;[7] a limiting energy within such a set is the asymptotic value corresponding to complete dissociation. A typical diagram[8] of these energies of such a system (in figure 1) thus resembles qualitatively an analogous diagram for a strongly bound molecular species such as CO. For the latter species CO specifically, near 300 K a vibrational state $v = 0$ is essentially the only one populated at thermal equilibrium, because the interval of energy $hc\omega_e$ to the first vibrationally excited state is $\approx 10 k_B T$ and the dissociation energy is a little less than $D_e \approx 430 k_B T$ at the dissociation limit relative to the energy at the separation of least energy.[9] Compared with CO that has about eighty bound vibrational (but rotationless) states, Ar_2 has about eight bound states below its limit corresponding to dissociation of the non-rotating molecule; as for Ar_2 at 300 K, $hc\omega_e \approx 1/8 k_B T$ and $D_e \approx 1/2 k_B T$,[8] gaseous argon exists in the form of diatomic molecules distributed over several bound states to an extent about one per cent of the total gaseous substance at a standard pressure 10^5 N m^{-2}. In their electronic ground states Ne_2 has about four bound rotationless states, whereas those of Kr_2 and Xe_2 number about 12 and 17 respectively.[8] The latter numbers are comparable with the number (14) of vibrational states of $^1H_2$,[10] consistent with the generally closer spacing of bound states for weakly bound than for strongly bound molecules of comparable reduced mass.

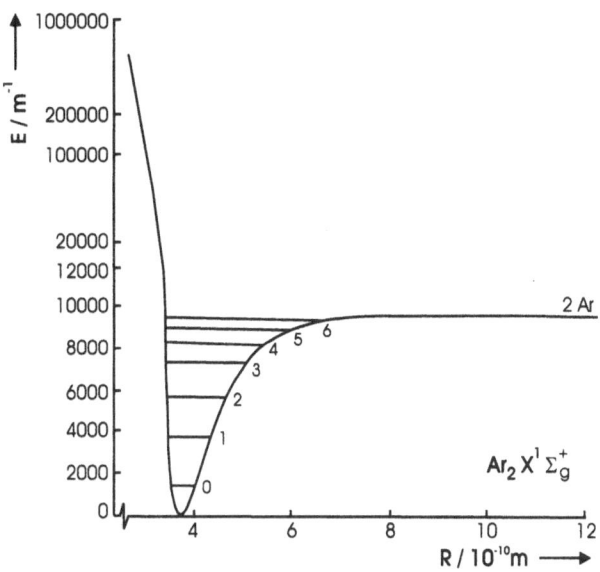

Figure 1. The potential-energy function of Ar_2 in the electronic ground state $X\ ^1\Sigma_g^+$ or O_g^+; the horizontal lines indicate the energies of predicted vibrational states but states near the dissociation limit are omitted for clarity; the ordinate scale is linear in the range $E < 1.3\ D_e$ and logarithmic thereabove; *reprinted from* J. Mol. Struct. 273, J.F. Ogilvie and F.Y.H. Wang, Potential-energy functions of diatomic molecules; I like nuclear species, p. 281 (1992), with kind permission of Elsevier Science–NL, Sara Burghartstraat 25, 1055 KV Amsterdam, The Netherlands.

Although for an attractive potential energy of a system in one dimension at least one vibrational state must exist,[11] for a real physical (or chemical) system in three dimensions no such requirement applies. Whether $^4He_2$ possesses a bound state (i.e. only $v = 0$) of the non-rotating molecule is controversial,[12,13,14] but certainly $^3He_2$ has none. Although for He_2 the distance R_e at a minimum energy of the function $V(R)$ is $2.75 \cdot 10^{-10}$ m, in a putative ground state $v = 0$ the average distance is $6.7 \cdot 10^{-9}$ m.[15] Hence in this case the value of a dynamic structural attribute differs greatly from that of an hypothetical static structure with no residual energy. Likewise for CO in the solar photosphere of which the temperature is ≈ 5800 K, due to a distribution over many vibrational states[16] the mean bond length is significantly greater than $R_e \approx 1.128 \cdot 10^{-10}$ m.

A conclusion of this consideration of properties of diatomic molecular species in gaseous samples is that any distinction between nominally intramolecular and intermolecular interactions must be arbitrary, based on some criterion of binding energy D_e or equilibrium internuclear distance R_e, or other; scaled appropriately, functions for potential energy of intramolecular and intermolecular interactions have a mutual resemblance. Many instances of diatomic molecular species in electronically excited states are known for which the binding energy is comparable with values for such weakly bound species as Xe_2.[9] As an instance of an analytic function for potential energy, that proposed by Morse[17] having a form

$$V(R) = D_e[1 - e^{-\alpha x}]^2 \tag{2}$$

51

in which $x = R/R_e - 1$, might serve as an approximate model for the true potential energy of such diverse species as non-polar and strongly bound H_2, polar and strongly bound HF and non-polar and weakly bound Ar_2, but for these three diatomic molecular species the parameters D_e, α and R_e assume widely varied values.[7]

On a basis of combined experimental evidence combined from macroscopic properties of noble gases, pure or in mixtures, these 'pair potentials' are deduced.[4] When such functions are converted into a suitable standard form, optical spectral properties of such systems are readily predicted, to serve as a guide to measurement of corresponding spectral transitions.[18] In this way microwave spectra of diatomic molecules in a supersonic jet, subjected to isentropic expansion to produce cooling near 10 K and thereby enhanced population of discrete vibrational and rotational states, could be observed.[19] Although data from scattering of molecular (really monatomic) beams of noble gases pertain formally to interactions of atoms only two at a time, a function for potential energy of NeXe in the bound region with E less than D_e deduced from such experiments[20,21] is much less accurate than a function based on bulk properties[22] in which 'many-body' interactions are naturally present, according to objective criteria of these microwave spectral lines;[19] the quality of data from such experiments on scattering is inferior to that from alternative measurements of intermolecular interactions. For liquid argon near its normal boiling point, interactions between atoms three (or more) at a time contribute at least about seven per cent of the cohesive energy of that condensed phase.[6]

3. BENZENE

We consider here a conventional molecular system for which chemical and physical data are abundant, even if necessarily incomplete. Although benzene C_6H_6 has a normal boiling point near 353 K, its melting point 278.7 K at 10^5 N m^{-2} much exceeds that of hexane C_6H_{14} at 178 K, or 1-hexene C_6H_{12} at 133 K, 2-hexene at 132 or 140 K, and 3-hexene at 135 or 160 K.[23] Solid benzene exhibits remarkable properties; for instance although the width of a line in the nuclear magnetic resonance spectrum decreases markedly near 95 K[24] and features in the infrared spectrum and a neutron diffractogram of a crystalline sample alter significantly near 230 K,[25] there seems to exist no strong crystallographic evidence of more than one structural phase in a range/K [4, 278] at 10^5 N m^{-2}. Evidence from nmr spectra was interpreted to imply near 95 K an onset of orientation, not free rotation, of molecular units about their hexad axes in the crystal lattice, but benzene seems not generally regarded to have a plastic-crystalline phase such as that of camphor or C_6F_{12}.[26] Substances of the latter type generally consist of globular molecules that reorient, subject to a barrier of energy that must be transcended, between equivalent conformations about almost any internal inertial axis, not one specific internal axis as in benzene, and diffusion of molecules in a crystalline phase is also generally more rapid than that in crystalline benzene.[26] Interactions between molecular units in benzene that allow retention of a lattice having a well defined structure, to a temperature large by comparison with kindred hydrocarbons, despite reorientation of these units, are somewhat mysterious. Such reorientation of an essentially rigid molecular unit about the hexad axis is a qualitative description of motion somewhere between a vibration involving a periodic angular displacement from an equilibrium conformation, or libration, and free or essentially unhindered rotation, which for a crystalline sample at normal pressure is exhibited uniquely by H_2 in its isotopic variants.[10]

A quantitative description of vibrations that leave unaltered the internuclear distances and interbond angles within each formula unit involves 21 optical branches for four molecular units per unit cell; nine of these are translational modes and twelve are rotational modes.[27] During each cycle of this rotational motion, the distances between nuclei of hydrogenic atomic centres in separate molecular units vary considerably (figure 2), typically

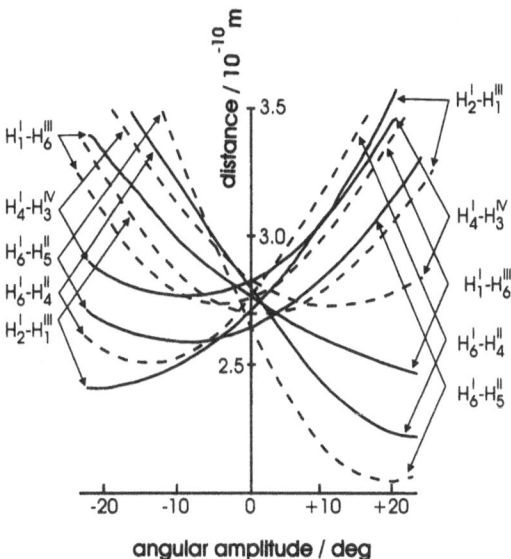

Figure 2. Variation of distances between hydrogen atomic centres during rotational motions of benzene molecules about their hexad axes; solid lines indicate that two specified molecules per unit cell execute the librational modes both in phase or both out of phase, whereas dashed lines indicate that two specified molecules execute their modes with one molecule in phase and another out of phase; *reprinted from* Spectrochim. Acta 22, M. Ito and G. Shigeoka, Raman spectra of benzene and benzene-d_6 crystals, p. 1039 (1966), with kind permission of Elsevier Science–NL, Sara Burghartstraat 25, 1055 KV Amsterdam, The Netherlands.

as much as $5 \cdot 10^{-11}$ m for vibrational modes that involve (root-mean-square) angular amplitudes/deg in a range [5, 8]. That wavenumbers of librational modes of either C_6H_6 or C_6D_6 show little, and only gradual, variation[28,29,30] with temperature/K in a range [4, 279] is taken as evidence against more than one phase existing at normal pressure.

In sum benzene thus fails to qualify as a mesocrystalline substance, in a sense of liquid crystal or plastic crystal.[26] Specimens having these mesophases provide superlative objects on which to investigate sensitive intermolecular interactions, because structural correlations are retained by nematic substances in one dimension, by smectic substances in two dimensions and by plastic-crystalline substances in three dimensions, although there is in each case much reorientational motion, and for liquid crystals also translational motion, despite retention of order over a large range. For liquid crystals there is a further variable in the shape of molecular units, prolate cylindrical or oblate cylindrical (discoid), in contrast with a generally globular shape of units in plastic crystals, and particular compounds pass through a sequence of as many as eight liquid-crystalline phases with temperature over a narrow range below a normal melting point.[26] Many experimental techniques are applicable to such investigation, some involving only simple (or inexpensive) equipment; results of such experiments not only provide fundamental information to elucidate the nature of intermolecular interactions but also can yield practical applications in various electronic devices.

Similarly to the phenomenon remarked for benzene, a marked decrease of width of a spectral line in an nmr experiment is observed for protons assigned to methyl groups at a terminus of long alkyl chains in unbranched carboxylic acids;[31] in this case an interpretation is that methyl groups begin to reorient about their local triad axis according to a small barrier of energy with respect to a hydrocarbyl chain that remains fairly immobile with respect to axes of a unit cell. This phenomenon, called pre-melting as it is observed not far below a normal melting point of a particular compound,[31] is a further manifestation of diverse intermolecular interactions in their dynamic aspects.

4. CRYSTALLINE SALTS

Other behaviour is observable for some inorganic compounds: we consider ZrW_2O_8 as a particular instance.[32]

Although a pseudo-binary phase diagram of the system $\{ZrO_2 + WO_3\}$ shows ZrW_2O_8 to be the only phase at pressure 10^5 N m^{-2}, the latter compound is stable thermodynamically only over a narrow range of temperature near 1373 K. At 298 K this structure has a cubic unit cell of space group $P2_13$; in this structure WO_4 tetrahedra and ZrO_6 octahedra are linked such that each octahedron shares corners with six separate WO_4 tetrahedra, but each WO_4 tetrahedron shares only three of its four oxygen ligands with adjacent octahedra. In this way one oxygen per tetrahedron appears bound solely to W; for this O (i.e. O4) on one W the next nearest adjacent metallic centre is distant about $3.6 \cdot 10^{-10}$ m. The length/10^{-10} m of a terminal W$-$O bond is significantly less than that of a bond for O between W and Zr, 1.705 versus 1.801 for W1 and 1.731 versus 1.783 for W2; the greater W$-$O distances are typical of those found in other compounds containing WO_4 tetrahedra. For a structural relationship between ZrW_2O_8 and some cubic vanadates and pyrophosphates AM_2O_7, in compounds of the latter type there is a single bridging O; both structures contain AO_6 octahedra and MO_4 tetrahedra linked by shared corners only, but an extra O in the Zr compound might be considered to split a unit $M_2O_7^{4-}$ into two WO_4^{2-} units. Although a distance $(2.3 \cdot 10^{-10}$ m) between O3 and W1 is too large to characterise a bond, there are nevertheless small distortions of interbond angles about W1 from an ideal tetrahedral value.

Experiments with diffraction of both xrays and neutrons indicate a structural phase transition to occur at 428 K, confirmed with dilatometric measurements.[32] A structure above this temperature is refined successfully only with a model comprising disordered W and O sites on a three-fold axis; the atomic disorder involves exchange of O3 between two tetrahedra. As a result of correlated thermal motion in the solid phase, a mean length of a bond can appear to decrease as temperature increases. For instance, for Zr$-$O,

$$R_{Zr-O}/10^{-10} \text{ m} = 2.0794 - 2.281 \cdot 10^{-5} \, T/K \tag{3}$$

Although an analogous 'shrinkage' is well established from experiments with diffraction of electrons from small polyatomic molecules in the gaseous phase, this effect is unlikely to be responsible for bulk properties that are peculiar to ZrW_2O_8.

Motion of another type involves rotation of rigid ZrO_6 octahedra about the axis [111]. A coupling of this rotation, via Zr$-$O$-$W linkages that are assumed to be flexible, results in marked contraction of the distance W1$-$O3; a negative angle of tilt corresponds to a counterclockwise rotation when viewed down [111]. A model involving translation, libration and a screwing motion indicates that when the angle of libration increases the dimension of a unit cell decreases. Transverse vibrations of Zr$-$O$-$W bonds involve coupled librations of the constituent polyhedra. Transverse thermal motions involving correlated rotation of polyhedra produce a net contraction of the cell dimension (figure 3). The net result is negative thermal expansion over temperature/K in the range [0, 1050]. The extent of contraction is large and almost constant over this range: the order of magnitude of this contraction for ZrW_2O_8,

$$\alpha \approx -9 \cdot 10^{-6} \text{ K}^{-1} \tag{4}$$

is comparable to the contrary effect measured for normal ceramics with large expansion,

$$\alpha \geq +8 \cdot 10^{-6} \text{ K}^{-1} \tag{5}$$

As this negative effect to a similar extent occurs on either side of a well defined transition between structural phases at 428 K, and as there is no significant variation of distance of a bond between metal and oxygen as a function of temperature, the latter effect fails to explain this uncommon property.

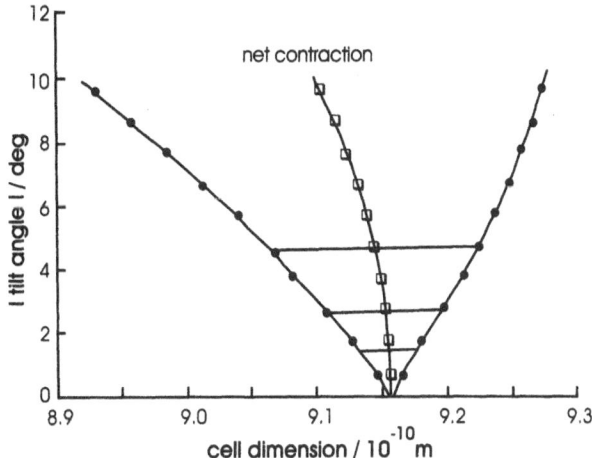

Figure 3. Magnitude of angle of polyhedral tilt versus dimension of the unit cell of ZrW_2O_8; the net cell parameter is shown as a function of polyhedral tilt; the figure depicts how harmonic libration of ZrO_6 octahedra produces a net contraction of length of the cubic cell; *reprinted with permission from reference 32, Copyright 1996 American Chemical Society.*

For comparison, water H_2O as ice in its normal hexagonal crystalline form I_h exhibits small thermal contraction below 73 K,[33] apart from density of the normal liquid phase increasing with temperature/K in a range [273.15, 277.13]; SiO_2 in three crystalline forms quartz, cristabolite and tridymite possesses this property but only above 1275 K.[34] A mechanism of correlated tilting in ZrW_2O_8, and analogously HfW_2O_8, provides a convincing explanation of thermal contraction of these crystals with temperature over a large range up to its point of decomposition, bestowing important practical applications on these materials.

5. CONCLUSION

We consider here aspects of intermolecular interaction of three prototypical chemical systems, progressing from diatomic molecules in the gaseous phase through molecular crystals to network crystals. In each case motions of molecular units (counting an atom of an element of group 18 as a monatomic molecule) in states of discrete energy produce effects on structural, spectral and other properties beyond what one might expect from a particular formula unit of a substance in its static structure. These dynamic interactions have ramifications for an understanding of not only chemical materials of which properties can be exploited in practical applications but also biological and physical processes; for this reason such dynamic interactions remain an important focus of scientific and technological investigation. Any modeling of these systems that fails to take into account such dynamic aspects of intermolecular interactions is likely to yield unreliable conclusions about properties of a material, but a method to generate values of parameters to model dynamic properties is not obvious.

REFERENCES

1. E.A. Moelwyn-Hughes, "Physical Chemistry," second edition, Pergamon, Oxford U.K. (1961).
2. F. London, *Z. Phys.* 63:245 (1930).
3. J.O. Hirschfelder, The nature of intermolecular forces, *in*: "Molecular Forces", J.O. Hirschfelder, ed., North-Holland, Amsterdam Holland (1967).
4. G.C. Maitland, M. Rigby, E.B. Smith and W.A. Wakeham, "Intermolecular Forces", Clarendon, Oxford U.K. (1981).
5. J. Lennard-Jones, *Proc. Roy. Soc. London* A112:214 (1926).
6. J.A. Barker, R.A. Fisher and R.O. Watts, *Mol. Phys.* 21:657 (1971).
7. J.F. Ogilvie, "Vibrational and Rotational Spectrometry of Diatomic Molecules", Academic, London U.K. (1998).
8. J.F. Ogilvie and F.Y.H. Wang, *J. Mol. Struct.* 273:277 (1992).
9. K.B. Huber and G. Herzberg, "Constants of Diatomic Molecules", van Nostrand Reinhold, New York U.S.A. (1979).
10. J. van Kranendonk, "Solid Hydrogen", Plenum, New York U.S.A. (1983)
11. J.B. Bronzan, *Am. J. Phys.* 55:54 (1987).
12. E.S. Meyer, J.C. Mester and I.F. Silvera, *J. Chem. Phys.* 100:4021 (1994).
13. F. Luo, G.C. McBane, G. Kim, C.F. Giese and W.R. Gentry, *J. Chem. Phys.* 100:4023 (1994).
14. W. Schöllkopf and J.P. Toennies, *J. Chem. Phys.* 104:1155 (1996).
15. F. Luo, C.F. Giese and W.R. Gentry, *J. Chem. Phys.* 104:1151 (1996).
16. C.B. Farmer, *Mikrochim. Acta* 3:189 (1987).
17. P.M. Morse, *Phys. Rev.* 34:438 (1929).
18. J.F. Ogilvie and F.Y.H. Wang, *J. Mol. Struct.* 291:313 (1993).
19. W. Jäger, Y. Xu and M.C.L. Gerry, *J. Chem. Phys.* 99:919 (1993).
20. C.Y. Ng, Y.T. Lee and J.A. Barker, *J. Chem. Phys.* 61:1996 (1974).
21. C.A. Linse, J. van den Biesen, E.H. van Veen and C.J.N. Meijenberg, *Physica* A99:166 (1979).
22. D.A. Barrow, M.J. Slaman and R.A. Aziz, *J. Chem. Phys.* 91:6348 (1989).
23. "Handbook of Chemistry and Physics", R.C. Weast, ed., CRC, Cleveland U.S.A. (1976).
24. E.R. Andrew and R.G. Eades, *Proc. Roy. Soc. London,* A218:537 (1953).
25. C.A. Swenson, W.B. Person, D.A. Dows and R.M. Hexter, *J. Chem. Phys.* 31:1324 (1959).
26. J.F. Ogilvie, *J. Chin. Chem. Soc.* 36:375 (1989); *J. Chin. Chem. Soc.* 36:501 (1989).
27. M. Ito and T. Shigeoka, *Spectrochim. Acta* 22:1029 (1966).
28. H. Bonadeo, M.P. Marzocchi, E. Castellucci and S. Califano, *J. Chem. Phys.* 57:4299 (1972).
29. M. Ghelfenstein and H. Szwarc, *Mol. Cryst. Liq. Cryst.* 14:273 (1971).
30. M. Ghelfenstein and H. Szwarc, *Chem. Phys. Lett.* 32:93 (1975).
31. M.R. Barr, B.A. Dunell and R.F. Grant, *Can. J. Chem.* 41:1188 (1963).
32. J.S.O. Evans, T.A. Mary, T. Vogt, M.A. Subramanian and A.W. Sleight, *Chem. Mater.* 8:2809 (1996).
33. K. Röttger, A. Endriss, J. Ihringer, S. Doyle and W. F. Kuhs, *Acta Cryst.* B50:644 (1994).
34. D. Taylor, *Brit. Ceram. Trans. J.* 83:129 (1984).

ATOMIC INTERACTIONS AND THE CHARGE DENSITY

Tibor Koritsánszky

Institute for Crystallography
Free University Berlin
Takustr. 6
14195 Berlin
Germany

1. INTRODUCTION

A rigorous physical description of interacting atoms or molecules requires the energy of the system to be given as a function of the nuclear configuration. In principle, quantum mechanics allows for a mapping of such a continuous energy–geometry relation but in practice, the energy can be derived only for a limited number of points of the nuclear configuration space. This can be done by decoupling the electronic and nuclear motion, i.e. by invoking the Born-Oppenheimer approximation which makes it possible to estimate the electronic energy through an approximate solution of the stationary Schrödinger equation at a given nuclear arrangement. It is important to recognize the difference between the molecular or cluster geometry, which is imposed as an initial condition on the equations of motion, and the structure, which emerges from the corresponding electronic wave function. The definition of the structure, in the "chemical" sense used here, includes the characterization of relevant interactions between chemical species in terms of atom–atom contributions, and thus, it requires atomic partitioning of the system considered. The theory of "atoms in molecules" has proven the distribution of electronic charge to contain sufficient information for deriving all the elements of the molecular structure. The physical basis for this resides in the dominant topological properties of the electron density. The quantum chemical derivation of this function is limited to relatively small molecular complexes in a hypothetical geometrical arrangement and allows one to study only short-range interactions. The experimental charge distribution, however, provides information on molecules in the crystalline state — in an energetically favored configuration supported by short, as well as long range interactions. The drawback of the experimental method has been its complicated and time-consuming nature. Due to the application of area detection in X-ray intensity collection, a major breakthrough is expected in the field and the method is becoming readily affordable.

The subject of the first part of this chapter is the distribution of charge and its fundamental topological properties based on which a molecular system can be characterized. The method of extracting the electron density from X-ray diffraction data is also outlined. This is followed by presenting examples for intermolecular interactions found in molecular crystals and characterized in terms of the topology of the experimental densities.

2. THE TOPOLOGY OF THE ELECTRON DENSITY

The electron density $\rho(\mathbf{r})$ can be obtained from the many-electron ground-state wave function by summing over the spin coordinates and integrating over the spatial coordinates of all the electrons but one ($\mathbf{r}$):

$$\rho(\mathbf{r}) = \sum_{spin} \langle \Psi | \Psi^* \rangle_{\mathbf{r}'} \tag{1}$$

If it is normalized to the total number of electrons it gives the probability for finding electronic charge in an infinitesimal volume element $d\mathbf{r}$. In the Born-Oppenheimer approximation[1] the molecular electron density is a parametric function of the actual spatial coordinates of the nuclei. For a single Slater determinant atomic wavefunction composed of molecular orbitals (Φ_i), $\rho(\mathbf{r})$ can be written as:

$$\rho = \sum_i n_i |\Phi_i|^2 = \sum_{ik} P_{ik} \varphi_i \varphi_k \tag{2}$$

where n_i's are the orbital occupation numbers (1 or 2), P_{ik} is the population matrix and φ_i are atomic orbitals, whose linear combinations gives rise to Φ.

Atomic interactions result in redistribution of electronic charge. That is why density deformations with reference to a "non-interacting" state appear to illuminate the basic characteristics of the interactions involved. Charge accumulations due to chemical bonds have been visualized, for a long time, in terms of the deformation electron density defined as the difference between the molecular and the superposition of isolated atomic densities ($\rho_{j,0}$):

$$\Delta\rho(\mathbf{r}) = \rho_{mol}(\mathbf{r}) - \sum_j \rho_{j,0}(|\mathbf{r}|) \tag{3}$$

This approach, in spite of its ability to deliver important information on the nature of charge rearrangements, remained a simple demonstration tool to draw only qualitative statements about chemical bonding.

The characterization of atomic interactions does not need a reference state. The necessary information emerges from the topology of $\rho(\mathbf{r})$ of the total system.[2] The determination of the regions in space where $\rho(\mathbf{r})$ decreases, increases and the location of its extrema requires the analysis of its gradient vector field ($\nabla\rho(\mathbf{r})$). This can be accomplished by examining the path portrait of the field, that is, by mapping the trajectories connecting the points of low density with those of high density. A path originates at infinity or at a point $\mathbf{r}_{min}$ where $\rho(\mathbf{r})$ exhibits a minimum, at least in one direction ($\nabla\rho(\mathbf{r}_{min}) = 0$ and one of the eigenvalues of the Hessian matrix is positive) and it terminates at a point $\mathbf{r}_{max}$ where $\rho(\mathbf{r})$ exhibits a maximum, at least in one direction ($\nabla\rho(\mathbf{r}_{max}) = 0$ and one of the principal curvature of $\rho(\mathbf{r})$ is negative). Two interacting atoms are connected via a maximum density path defined by two trajectories of $\nabla\rho(\mathbf{r})$. Both originate somewhere midway between the nuclei — at the bond critical point (BCP), where $\rho(\mathbf{r})$ attains it minimum value — and terminate at each of the nuclei, where $\rho(\mathbf{r})$ exhibits its local maxima. The presence of such a bond path (BP) between a pair of nuclei is a necessary and sufficient condition for the existence of an interaction between the atoms. The network of the bond paths defines the molecular graph. The trajectories terminating at the BCP span a surface, the interatomic surface, in which $\rho(\mathbf{r})$ attains its maximum value at the BCP. Here the two principal curvatures are negative and the associated eigenvectors of the Hessian matrix define a plane perpendicular to the bond path.

To each nucleus belongs a neighborhood (atomic basin) satisfying the condition that any trajectory originating in it terminates at the nucleus. The interatomic surface provides a boundary between the basins of the neighboring atoms. The topological analysis of the electron density leads to the reconstruction all elements of the molecular structure and also to the characterization of atomic interactions and chemical bonds. The length of the BP, the location of the BCP, $\rho(\mathbf{r})$ and its principal curvatures at the BCP $(\mathbf{r}_b)$ can be used for this purpose. The latter quantities are a sensitive figure of the strength of a bond. From $\rho(\mathbf{r}_b)$ an expression for the bond order can be derived.[3] For homopolar bonds the ratio of the two negative principal curvatures of $\rho(\mathbf{r})$ at the BCP, a quantity giving the extent of departure from cylindrical symmetry, reflects the π character of bond of the bond.[4]

The trace of the Hessian matrix $(\nabla^2\rho(\mathbf{r}))$, known as the Laplacian, is negative or positive depending on whether electronic charge is locally concentrated or depleted. Its topology is characteristic of a given atom. The region of the outer shell of an atom over which the Laplacian is negative is called the valence shell charge concentration (VSCC). The topology of the VSCC, given by its maxima in number and location, is in accordance with the Lewis and valence shell electron pair repulsion models.[5] To each local maximum in the VSCC a pair of bonded or nonbonded electrons can be assigned.

Maxima in the VSCC of the atoms along the bonds are called bonded charge concentrations. Additional maxima in the VSCC (non-bonded charge concentrations) correspond to the lone-pair electrons. The concentration of electronic charge in the interatomic surface (relatively large values of $\rho(\mathbf{r})$ and the negative Laplacian at the BCP) is typical for shared interactions (covalent bonds). In the case of closed-shell interactions the charge is contracted towards the nuclei leading to a region of positive Laplacian between them. The most intense local charge concentrations/depletions in a molecule are the regions where nucleophilic/electrophilic attacks are most likely to take place.[6]

Figure 1. Map of the gradient vector field of the theoretical electron density (HF/6-311++($3df, 3pd$)) of 1,2,4-triazole cluster in the plane of the N−H···N hydrogen bridge bond.

As an example, the theoretical gradient vector field of $\rho(\mathbf{r})$ is displayed in Figure 1 in terms of a trajectory map in the plane of an N−H···N hydrogen bridge in the 1,2,4-triazole dimer.[7] The three atoms are connected via two BPs, the shorter one corresponds to the N−H covalent bond whereas the longer one is a sign for the H···N interaction. The surfaces defined by the trajectories terminating at the BCPs enclose the central hydrogen atom. The near planarity of the interatomic surface between the atoms forming the H···N contact is typical of hydrogen bonds. For the N–H shared interaction charge is accumulated in the surface and contracted towards the bond path, while for the H···N closed-shell interaction charge is removed from the surface and concentrated in the basins of the participating atoms.

3. EXPERIMENTAL ELECTRON DENSITY

The procedure of experimental charge density determination involves the interpretation of Bragg intensities:[8]

$$I_{Bragg} = |F(\mathbf{H})|^2 \tag{4}$$

where $F(\mathbf{H})$ is the structure factor — the Fourier transform of the thermally averaged density $\langle \rho(\mathbf{r}) \rangle$,

$$F(\mathbf{H}) = \int_V \langle \rho(\mathbf{r}) \rangle \exp(i\mathbf{H}'\mathbf{r}) d\mathbf{r} \tag{5}$$

and $\mathbf{H}$ is the Bragg vector with integral components h, k, l, relative to the reciprocal axes $\mathbf{a}^*, \mathbf{b}^*, \mathbf{c}^*$.

The goal is to extract an analytical static density $\rho(\mathbf{r})$ from a set of measured $|F(\mathbf{H})|$ data. This requires the deconvolution of thermal motion, that is, a model for describing the effect of the nuclear motion on the density. It is assumed that $\rho(\mathbf{r})$ can be expressed in terms of pseudoatomic densities (ρ_k):

$$\rho(\mathbf{r}) = \sum_k \rho_k(\mathbf{r} - \mathbf{r}_k) \tag{6}$$

each of which is convoluted with the probability distribution function describing the motion of the kth nucleus in thermal equilibrium:

$$P_k(\mathbf{u}) = (2\pi)^{-\frac{3}{2}} (\det \mathbf{U_k})^{-\frac{1}{2}} \exp\left(-\frac{1}{2}\mathbf{u}'\mathbf{U_k}^{-1}\mathbf{u}\right) \tag{7}$$

where $\mathbf{u}$ is the displacement vector of the kth nucleus from its equilibrium position and $\mathbf{U}_k$ denotes the corresponding mean-square displacement amplitude tensor. In the aspherical-atom formalism developed by Hansen and Coppens[9] the density units are composed of three parts:

$$\rho(\mathbf{r}) = \rho_c(r) + P_v \rho_v(\kappa r) + \rho_d(\kappa' r) \tag{8}$$

where ρ_c and ρ_v are the spherical core and valence densities, respectively, and

$$\rho_d(\kappa' r) = \sum_l R_l(\kappa' r) \sum_m P_{lm} y_{lm}(\mathbf{r}/r) \tag{9}$$

is the term which accounts for valence deformations. The radial functions $R_l(r)$ are taken as simple Slater functions[10] (modified by the screening constants, κ'), while y_{lm} are density normalized real spherical harmonics. Beside the conventional parameters, the P_v, P_{lm}, κ and κ' are the variables of the least-squares procedure and they are obtained by fitting the structure factor amplitudes predicted by the model to those measured.

The angular dependence of the valence deformation density is described in local frames centered at each atomic site which makes it possible to impose local symmetry in a convenient way. The number of parameters to be refined can be reduced also by assuming chemical equivalence between atoms involved in similar bond systems.

The recovering of the density from its three dimensional Fourier image relies on assumptions directly not deducible from the data, consequently, it needs support from theory. The density obtained does not correspond to a quantum state, i.e. to a wave function. If the model applied is unambiguous the image fitted to the intensity data can be assumed to be topologically equivalent to the exact density, but there is no guarantee for it. In spite of this fact numerous studies have demonstrated that there is a one-to-one mapping between experimental and theoretical densities.[11] This homomorphy is revealed by the same number and type of CPs found in the intramolecular regions of the molecular densities obtained from diffraction data and from an approximate wave function of the isolated system. There are certain strict experimental requirements that limit the application of the diffraction method. It is essential to maintain kinematical conditions, as much as possible, during the entire measurement. The data should include precise Bragg intensities, also at high scattering angles ($\sin\theta/\lambda > 1.0$ Å^{-1}) which can be achieved only at low-temperature ($T \sim 100$ K or even lower). To reveal systematic errors, due to dynamic scattering, symmetry equivalent reflections have to be collected. Such experiments have been based on serial measurements of intensity data performed with single-reflection counters. Due to the application of area detection in X-ray intensity collection the field is being revolutionized. This technique opens a completely new perspective for studies of larger molecules.[12] In such cases, although the number of reflections increases drastically, there is almost no increase in the measuring time, if an area detector is used.

4. APPLICATIONS TO MOLECULAR CRYSTALS

A crystal is a supermolecule. Its structure is revealed from the topology of the density of the crystalline state, in the same way as the molecular structure is revealed from the topology of the molecular charge density $\rho(\mathbf{r})$. Atoms in a crystal can be identified with bounded regions of real space. The surfaces enclosing an atomic basin in a crystal satisfy the zero-flux condition, just as the interatomic surfaces do in a molecule. The topological analysis of the crystalline density is of special importance in the study of intermolecular interactions. It gives a clear answer to the basic questions; how to identify the dominant forces of crystal formation, and how to decompose them into atomic interactions. Given the density of the unit cell, all CPs can be located and all bond paths can be mapped. In this section applications of the formalism described above to some molecular crystals are given.[13] Based on the static $\rho(\mathbf{r})$ extracted from high-resolution X-ray diffraction data the topological indices of typical atom-atom interactions are examined and compared to those obtained from quantum-chemical calculations[14] of the isolated molecules.

The Hydrogen Bond System of 1,2,4-triazole

An interesting structural aspect of this five-membered heterocycle[7] is the asymmetry reflected in the ring's geometry. The neutron data[15] show that the two formal single bonds are not equivalent; the C(3)−N(4) bond (1.3656(7) Å) is longer than that formed by the C(5) and N(1) (1.3345(7) Å) atoms (Figure 3). The latter one is practically equal in length to the C(5)−N(4) (1.3343(7) Å) double bond. This geometry suggests aromaticity but it is not supported by theoretical calculations on the isolated molecule. The bond distances, in the above order, are 1.3618, 1.3539 and 1.3286 Å for the optimized geometry at the MP2/6-311+(2d,2p) level. The question raises if the asymmetry described is the result of intermolecular interactions.

The molecules in the crystal form corrugated sheets via relatively strong N−H⋯N hydrogen bonds with atoms N(1) and N(4) being donors and acceptors, respectively (Figure 2). This arrangement, as depicted in Figure 3, can formally be represented by a structure (B) which is intermediate between the two tautomers (A, C) and in which the single and double bonds

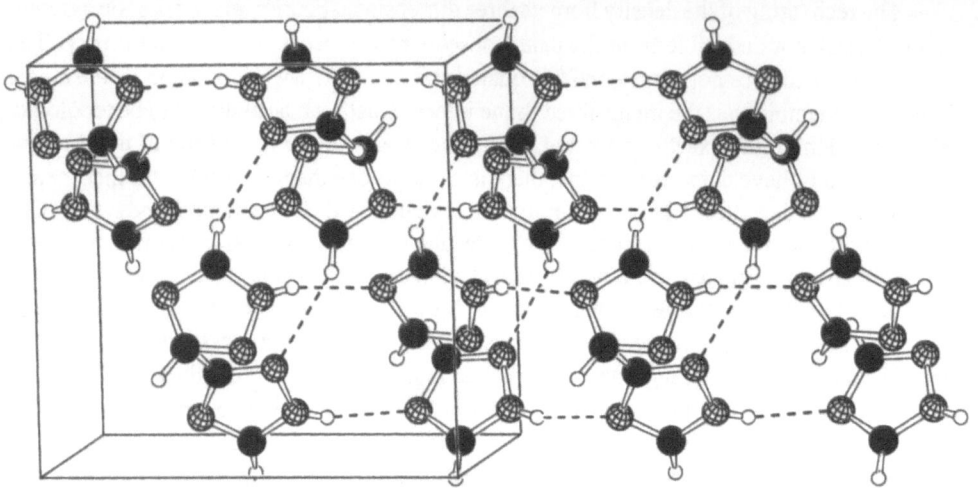

Figure 2. A crystal packing diagram of 1,2,4-triazole including the hydrogen bond system along the **a** axis. The carbon atoms are represented by black circles.

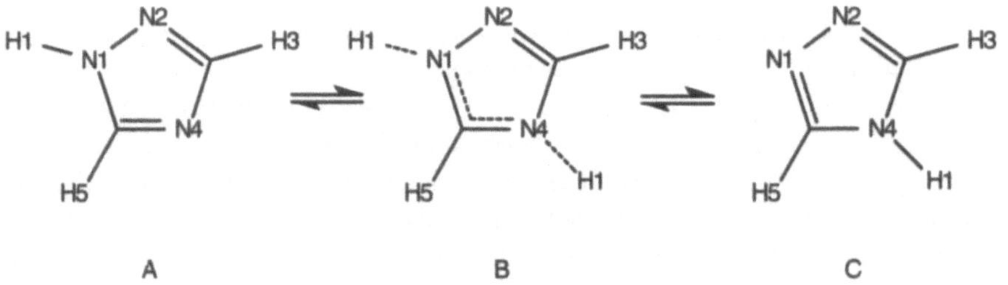

Figure 3. Resonant forms of 1,2,4-triazole.

become equivalent. This simple picture suggests that the intermolecular interactions can be considered as the source of balancing the two bonds in question. At the HF/6-311++($3df$,$3pd$) level single point calculations were performed not only on the isolated molecule but also on a cluster of three molecules arranged as in the crystal to mimic the effect of the neighbors on $\rho(\mathbf{r})$. Significant changes can be found for the Laplacian in the bonds to the N(4) atom and in that of N(1)−H(1). An important result is the slight increase of charge concentration in the formal single C(5)−N(1) bond in the cluster with respect to the values found in the isolated molecule. The density at the BCP increases from 2.30 to 2.35 $e/\text{Å}^3$, while the Laplacian at the same point decreases from −25.4 to −28.7 $e/\text{Å}^5$, as a result of including the neighbors in the calculation. This effect is well represented by the difference density contour diagram displayed in Figure 4. This map was obtained by subtracting the density of the isolated molecule from that of the middle one in the cluster of three. It shows charge rearrangements due to the polarization of $\rho(\mathbf{r})$ at each atomic site. As a result, charge accumulations occur in the N(1)−H(1) and N(1)−C(5) bonds. The experimental bond topological indices for the two bonds to C(5) were found to be equal ($\rho(r_b) = 2.27$ and 2.26 $e/Å^3$, $\nabla\rho(r_b) = -21.2$ and -19.5 $e/Å^5$, respectively, for the C(5)−N(4) and the C(5)−N(1) bonds). The BCP of the H(1)···N(4) hydrogen bond occurs at a distance of 0.558 Å from the hydrogen atom toward the acceptor. The density is low (0.20 $e/Å^3$), while the Laplacian is positive (1.4 $e/Å^5$) at this site. Both values are characteristic of closed shell interactions.

Figure 4. Difference of the electron density of the isolated and the middle over hydrogen bridge bonds connected molecule, based on the theoretical calculations (HF/6-311++(3df,3pd)). Contour lines are drawn at 0.01, 0.03, 0.05, 0.1, 0.3 and 0.5 $e/Å^3$. Contours in negative regions are dashed.

Hydrogen Bonds in the Crystalline D,L-Aspartic Acid and L-Threonine

In the course of our study on the transferability and reliability of the experimental $\rho(r)$ we have been analysing the topology of different covalent bonds, as well as non-bonded interactions, in chemically analogous molecules, such as the natural aminoacids.[16] The results presented here are based on "state of the art" measurements, performed at 20 K using AgKα radiation. The X-ray data of D,L-aspartic acid and L-threonine are the most extended ($\sin(\theta)/\lambda = 1.368$ and 1.346 Å^{-1}, respectively) ever collected with conventional technique. Both molecules, found in their zwitterionic form in the crystal, form complicated hydrogen bonding systems in which all hydrogen atoms of the NH$_3^+$ groups are donors. In both cases, one of the oxygen atom of the carboxylate group is involved in two, whereas the other one, only in one intermolecular hydrogen bonds. In addition, there are oxygen atoms present in both crystals which bond to C−H donors with O···H distances shorter than the sum of the van der Waals radii (2.6 Å). The bonding situation in both crystals are depicted in Figure 5.

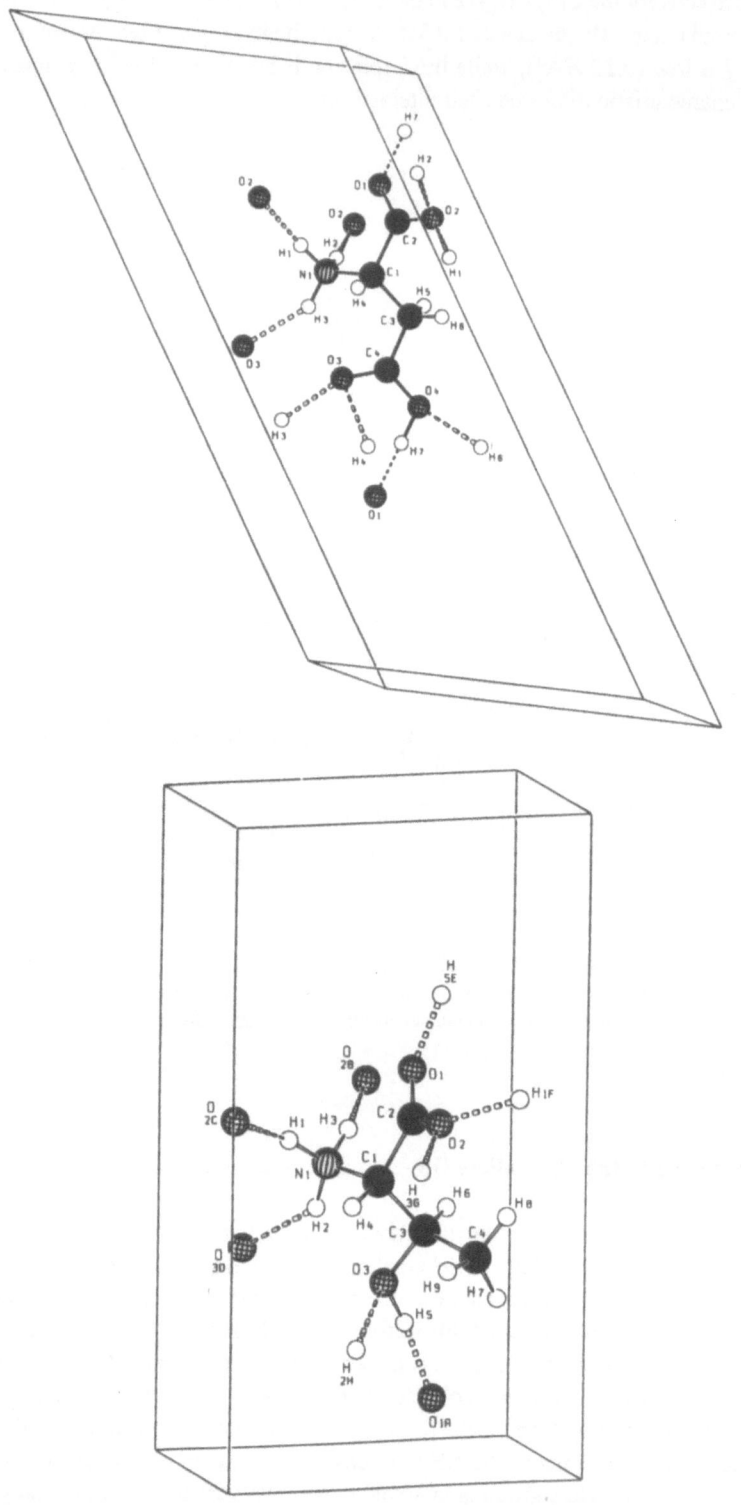

Figure 5. The hydrogen bond systems in the crystal of D,L-aspartic acid (top) and L-threonine (bottom).

Table 1. Bond Topological Indices of O$\cdots$H Hydrogen Bonds in D,L-Aspartic Acid and L-Threonine[a]

O$\cdots$H	symm. op./transl.	R(A$\cdots$H)	α(A$\cdots$H–D)	ρ	$\nabla^2\rho$	R(A–CP)	δ
		DL-Aspartic Acid					
O(1)$\cdots$H(7)	$1/2+x, 1/2-y, 1/2+z$ 0 0 0	1.5222(2)	178.6(1)	0.50(1)	1.1(2)	1.050	0.014
O(2)$\cdots$H(1)	$x, -y, 1/2+z$ 0 1 0	1.8075(2)	175.5(1)	0.25(1)	3.6(2)	1.153	0.012
O(2)$\cdots$H(2)	$1/2-x, 1/2+y, 1/2-z$ 1 0 1	1.8480(2)	172.7	0.26(1)	3.2(2)	1.159	0.012
O(3)$\cdots$H(3)	$-x, y, 1/2-z$ 1 0 0	1.9794(2)	148.3	0.10(1)	2.2(2)	1.276	0.219
O(3)$\cdots$H(4)	$-x, -y, -z$ 1 1 1	2.5033(2)	134.9	0.04(1)	0.8(2)	1.489	0.213
O(4)$\cdots$H(6)	$-x, y, 1/2-z$ 1 0 1	2.5621(2)	127.6(1)	0.04(1)	0.7(2)	1.469	0.141
		L-Threonine					
O(1)$\cdots$H(5)	$1/2+x, 1/2-y, -z$ 0 1 0	1.7278(2)	161.9(1)	0.28(1)	2.67(1)	1.142	0.0022
O(2)$\cdots$H(1)	x, y, z 0 0 -1	1.8006(3)	157.5(1)	0.15(1)	3.88(1)	1.174	0.0084
O(2)$\cdots$H(3)	$-x, 1/2+y, 1/2-z$ 1 -1 0	1.8640(3)	161.1(1)	0.09(1)	1.91(1)	1.233	0.0780
O(3)$\cdots$H(2)	$1/2-x, -y, 1/2+z$ 1 1 -1	2.1919(3)	138.8(1)	0.07(1)	1.18(1)	1.347	0.0561
O(3)$\cdots$H(4)	$1/2-x, -y, 1/2+z$ 1 1 0	2.2125(3)	158.0(1)	0.06(1)	1.44(1)	1.369	0.0225
O(1)$\cdots$H(1)	$-x, 1/2+y, 1/2-z$ 1 -1 -1	2.5126(3)	108.0(1)	0.06(1)	0.89(1)	1.408	0.0804

[a]Units are in e, Å and degrees. R(A$\cdots$H) and R(A–CP) are the distances of the acceptor atoms from the hydrogen atom and from the CP of the A$\cdots$H interaction, respectively. δ denotes the difference between the bond path length and the interatomic distance, α(A$\cdots$H–D) is the angle formed by the acceptor, hydrogen and donor atoms.

The analysis of the non-bonded interactions in terms of the topology of the experimental $\rho(\mathbf{r})$ delivered chemically significant results. All intermolecular first-neighbor atom-atom connections were examined but BCPs were located only for the O···H−X (X = O, N, C) hydrogen bonds described above. In Table 1 the geometrical and topological parameters of these interactions are listed for both crystal structures. The values, as expected, suggest closed shell interactions. The stronger the bond the higher the density located between the O and H atoms. For the strong (weak) hydrogen bonds the interaction path length is slightly (considerably) longer than the geometrical length of the interaction line connecting the donor with the acceptor.

Covalent and Hypervalent S−O Bonds *versus* S···O Interactions

The description of the bonding in molecules which apparently violate the octet rule has invoked the concept of hypervalency.[17] A tetracoordinate sulfur compound, for example, possesses a regular or distorted trigonal bipyramidal geometry in which the axial ligands form a linear arrangement with the central sulfur atom with distances significantly longer than the usual single bonds. This bonding is called hypervalent and described as a three-center four-electron interaction which leads to a smaller bond order than that encountered in the corresponding two-electron two-center system. In homosubstituted sulfuranes the axial bonds are longer than the equatorial bonds, while in an asymmetrical X−S−Y arrangement the bond polarizations and the bond distances are controlled by the electronegativity of the axial ligands.

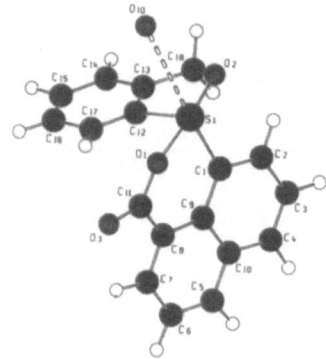

Figure 6. Molecular drawing of diaryl(alkoxy)(acyloxy)spiro-λ^4-sulfane including the oxygen atom of the dioxane solvent coordinated to the sulfur.

Structural studies on representative sulfuranes and sulfonium salts have successfully characterized almost the whole range of S−O interactions in O−S−O systems including covalent, hypervalent bonds and nonbonded interactions.[18] Equal hypervalent S−O bonds exist only in symmetric molecules such as the diaryldiacyloxyspirosulfuranes where the hypervalent S−O axial bond is 1.83 Å long compared to the usual covalent bond length of 1.70 Å. If one of the S−O bond is polarized the other gains substantial covalent character, i.e. the lengthening of the former bond is accompanied by the shortening of the other one. In an extreme case the three-center system splits into a covalent bond and an S···O close contact. The diaryl(alkoxy)(acyloxy)spiro-λ^4-sulfane,[19], seen in Figure 6, represents an intermediate unsymmetric O−S−O system. The unit cell contains also a half molecule of dioxane, the oxygen atom of which (O(1d)) is connected to the sulfur atom. The O(1d) atom is in the equatorial plane (C(12)−S(1)−C(1) plane) and the S(1)···O(d) interaction line coincides with the S(1)−C(1) bond.

The topological analysis of the experimental density has led to the BCPs and bond indices summarized in Table 2. Here all interactions in which the central sulfur atom is involved are characterized. For the intramolecular S−X bonds the theoretical results, at HF/6-311**(2df,2p) level, are also given. The difference between the two S−O bonds is clearly seen in terms of the $\rho(r_b)$ and $\nabla\rho(r_b)$ values. In the stronger bond (S(1)−O(2)) the density is higher than in the weaker bonds (S(1)−O(1) and S(1)···O(1d)), the Laplacian is positive at all of the S−O BCPs, except for the (S(1)−O(2)) bond. The density is the highest at the BCPs of the S−C covalent bonds where a continuous region of negative Laplacian can be found.

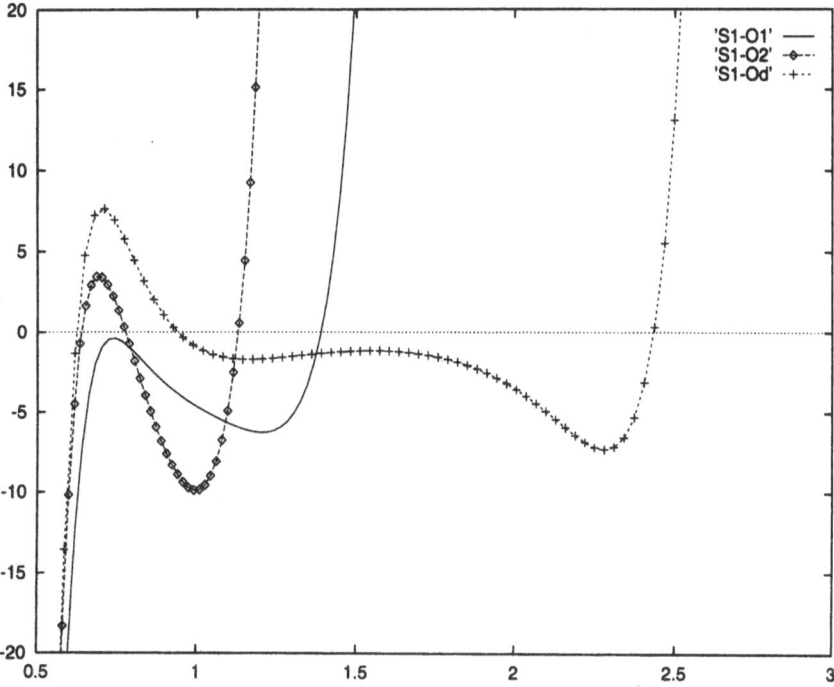

Figure 7. The variation of the Laplacian function along the bond paths of the S-O interactions in diaryl(alkoxy)(acyloxy)spiro-λ^4-sulfane.

Table 2. Topological parameters of bonds to the sulfur atom in diaryl(alkoxy)(acyloxy)spiro-λ^4-sulfane[a]

Bond	ρ	$\nabla^2\rho$	R_{S-X}	R_{S-BC}
S(1)−O(2)	1.26(1)	−0.5(1)	1.694(2)	0.787
	1.23	−1.5	1.694	0.711
S(1)−O(1)	0.60(1)	4.2(1)	2.045(2)	0.983
	0.54	3.7	2.045	1.044
S(1)−O(1d)	0.1(1)	1.2(1)	3.054(2)	1.628
S(1)−C(1)	1.36(1)	−2.6(1)	1.825(1)	0.969
	1.32	−9.6	1.825	1.033
S(1)−C(12)	1.39(1)	−2.0(1)	1.811(1)	0.964
	1.35	−9.9	1.811	0.920

[a] The second entry for each bond refers to values obtained from the HF/6-311G** wave function of the isolated molecule. Units are in e and Å.

Figure 7 displays the variation of $-\nabla\rho(\mathbf{r}_b)$ along the bond path of the S–O bonds discussed. The peaks situated about 0.7 Å away from the sulfur atom correspond to its VSCC. There is a bonded VSCC in the direction of the S(1)–O(2) covalent bond but no charge accumulation is observed in the valence shell in the direction of the S(1)–O(1) "hypervalent" bond. The largest peak found along the S(1)$\cdots$O(d) interaction line corresponds to the nonbonded VSCC of the sulfur atom.

Van der Waals Interactions

The bond paths found between atoms of neighboring molecules in a crystal indicate atom-atom interactions being relevant to maintaining a given packing. In molecular crystals these interaction lines can form quite a complicated and sometimes unexpected three dimensional network. An interesting example is the crystal structure of pentafluorosulfanyl isocyanate (F_5SNCO). The X-ray diffraction data were collected at 130 K using a single crystal grown in situ in a capillary on the diffractometer. In the space group $P\bar{1}$ the asymmetric unit consists of two molecules with different conformations; the NCO group is either staggered (s) or eclipsed (e) with respect to the closest equatorial fluorine atom in the SF_4 plane. The peripheral atoms in these molecules are F and O atoms, thus one could expect only F$\cdots$F, O$\cdots$O and F$\cdots$O type interactions. The crystal packing is shown in Figure 8. The representation on the left is a projection onto the bc crystallographic plane showing two layers which are situated parallel to the c axis and formed by an alternated sequence of the different conformers. There is another layer parallel to the b axis formed by the eclipsed conformers (right figure). In this layer the isocyanate groups are in a head-to-head orientation in the c direction. The topological analysis of the experimental static $\rho(\mathbf{r})$ led to the location of three intermolecular BCPs around each oxygen atom. The corresponding BPs define the O$\cdots$O interaction lines. The oxygen atoms, connected in such a way, form a network of six-membered rings of chair conformation. In Figure 8 two kinds of rings can be recognized: one of them is characterized by the O(s)–O(s)–O(e)–O(s)–O(s)–O(e), while the other one, by the O(e)–O(e)–O(s)–O(e)–O(e)–O(s) sequence, where O(e) and O(s) refers to oxygen atoms in the eclipsed and staggered conformers, respectively. There are four symmetry independent O$\cdots$O bond paths. The geometrical and topological parameters of the corresponding contacts are listed in Table 3.

Table 3. Experimental Geometrical and Topological Parameters of O$\cdots$O Interactions in the Crystal of F_5SNCO[a]

at1$\cdots$at2	symm. op. for at2	R(at1$\cdots$at2)	ρ	$\nabla^2\rho$	R(at1–CP)	R(CP-at2)
O(s)$\cdots$O(s)	$1-x, 1-y, -z$	3.312(3)	0.024(1)	0.360(1)	1.660	1.660
O(e)$\cdots$O(s)	$-x, 2-y, -z$	3.251(2)	0.023(1)	0.406(1)	1.486	1.754
O(e)$\cdots$O(s)	$1-x, 2-y, -z$	3.228(2)	0.023(1)	0.405(1)	1.736	1.504
O(e)$\cdots$O(e)	$-x, 2-y, -z$	3.103(3)	0.033(1)	0.534(1)	1.550	1.550

[a]Units are in e and Å. Label s (e) stands for oxygen atoms in the staggered (eclipsed) conformer.

Based on these entries the O(e)$\cdots$O(e) interaction is found to be the strongest, the O(s)$\cdots$O(s) the weakest, while those of O(s)$\cdots$O(e) type are of intermediate strength. The

O−BCP distances are between 1.550 and 1.660 Å which are significantly larger than the van der Waals radius of the oxygen atom (1.4 Å).

The F···F interactions form a less symmetrical and much more complicated pattern. The intermolecular BCPs corresponding to the F···F contacts were found between 1.50 and 1.60 Å away from the fluorine atoms. The $\rho(\mathbf{r_b})$ values are smaller than those found for the O···O interactions (0.03–0.02 e/Å^3). It is important to mention that no BCPs corresponding to a F···O type interactions were located.

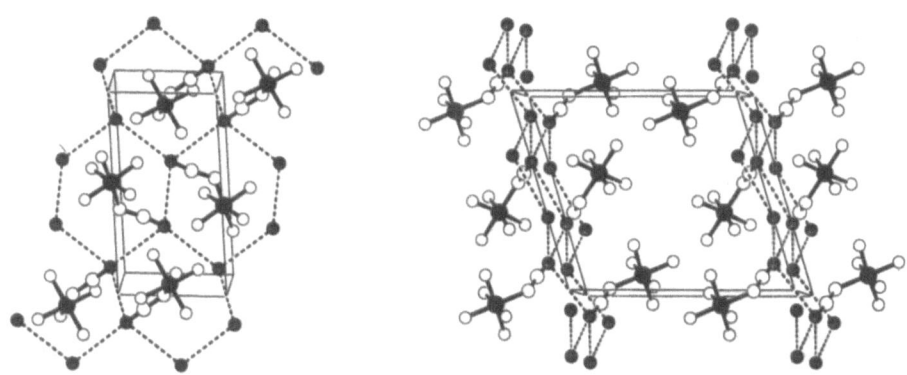

Figure 8. Graphical representations of the crystal packing of pentafluorosulfanyl isocyanate; a projection onto the **bc** crystallographic plae (left), a layer of the eclipsed conformers parallel to the **b** axis (right).

Acknowledgment

The results presented here are based on the X-ray data collected by Dr. J. Buschmann and Dr. D. Zobel. Their contribution is gratefully acknowledged. I also thank to R. Flaig for converting the text into LATEX format.

REFERENCES

1. M. Born and R. Oppenheimer, *Ann. Phys.* 84:457 (1927).
2. R.F.W. Bader. "Atoms in Molecules", Oxford Science Publications, Clarendon Press, London (1990).
3. R.F.W. Bader, T.S. Slee, D. Cremer, and E. Kraka, *J. Am. Chem. Soc.* 105:5061 (1983).
4. D. Cremer, E. Kraka, T.S. Slee, R.F.W. Bader, C.D.H. Lau, T.T. Nguyen-Dang and P.J. MacDougall, *J. Am. Chem. Soc.* 105:5069 (1983).
5. R.J. Gillespie, I. Bytheway, R.S. DeWitte and R.F.W. Bader, *Inorg. Chem.* 33:2115 (1994).
6. R.J. Gillespie, R. Kapoor, R. Faggiani, C.J.L. Lock, M. Murchie and J. Passmore, *J. Chem. Soc. Commun.* 8:1 (1983).
7. P. Fuhrmann, T. Koritsánszky and P. Luger, *Zeitschrift für Krist.* 212:213 (1995).
8. R.F. Stewart and D. Feil, *Acta Cryst.* A36:503 (1980).
9. N.K. Hansen and P. Coppens, *Acta Cryst.* A34:909 (1978).
10. E. Clementi and C. Roetti, *Atomic Data and Nuclear Data Tables* 14:177 (1974).
11. C. Gatti, R. Bianchi, R. Destro and F. Merati, *J. Mol. Struc. (Theochem)* 255:409 (1992). C. Flensburg, S. Larsen, R.F. Stewart, *J. Phys. Chem.* 99:10130 (1995). T. Koritsánszky, J. Buschmann and P. Luger, *J. Phys. Chem.* 100:10547 (1996).
12. T. Koritsánszky, R. Flaig, D. Zobel, , H.-G. Krane, W. Morgenroth and P. Luger, *Science* (1997), in press.
13. T. Koritsánszky, S. Howard, R.P. Mallinson, Z. Su, T. Richter, N.K. Hansen: (1995). XD, A Computer Program Package for Multipole Refinement and Analysis of Charge Densities from X-ray Diffraction Data. *User Manual* FU-Berlin (1995).

14. M.J. Frisch, G.W. Trucks, M. Head-Gordon, P.M.W. Gill, M.W. Wong, J.B. Foresman, B.G. Johnson, H.B. Schlegel, M.A. Robb, E.S. Replogle, R. Gomperts, J.L. Andres, K. Raghavachari, J.S. Binkley, C. Gonzalez, R.L. Martin, D.J. Fox, D.J. Defrees, J. Baker, J.J.P. Stewart and J.A. Pople, Gaussian 92, Revision C, Gaussian, Inc., Pittsburgh PA (1992).
15. G.A. Jeffrey, J.R. Ruble and J.R. Yates, *Acta Cryst.* B39:388 (1983).
16. R. Flaig, T. Koritsánszky, D. Zobel and P. Luger, *J. Am. Chem. Soc.* (1997), in press.
17. J.I. Musher, *Angew. Chem. Int. Ed. Engl.* 8:54 (1969).
18. A. Kálmán, *Croatica Chemica Acta* 66:519 (1993).
19. D. Szabó, I. Kapovits, Gy. Argay, M. Czugler, A. Kálmán and T. Koritsánszky, *J. Chem. Soc. Perkin. Trans.* 2:1045 (1997).

CYCLOMETALLATION OF ALKYLPHOSPHINES

Michael T. Benson[1] and Thomas R. Cundari[2]

[1]Department of Chemistry
University of Michigan
Ann Arbor, MI 48109-1055

[2]Department of Chemistry
The University of Memphis
Memphis, TN 38152-6060

1. INTRODUCTION

Catalytic activation of alkane C–H bonds is of great technological importance in the efficient utilization of petrochemical feedstocks.[1] Phosphines are widely used in many catalysts as co-ligands to control a wide variety of operational variables such as solubility in different solvents, stereospecificity, and regiospecificity.[2] A very large number of chiral and achiral phosphines, of differing denticity, have been synthesized and employed by organometallic chemists. A major obstacle to more widespread use of phosphine complexes in industrial catalysis arises from their degradation by oxidative addition pathways resulting from P–C bond cleavage and cyclometallation.[3]

Intramolecular C–H activation by the 14-electron model $Ir(PH_3)(PH_2R)(H)$, (where R = methyl, ethyl, propyl, eq. 1) is presented. This research is an extension of methane and ethane activation[4,5] using a model of an alkane dehydrogenation catalyst studied experimentally by Crabtree.[6,7]

$$Ir(H)(PH_3)(\eta^1\text{-}PH_2R) \rightarrow Ir(H)_2(PH_3)(\eta^2\text{-}R'PH_2) \tag{1}$$

The putative intermediate is $Ir(\eta^1\text{-carboxylate})(PR_3)_2$, a 14-electron species. Methane activation is intermolecular, while ethane activation involves both inter- and intramolecular processes,[4,5] making it of interest to compare them with cyclometallation.

(2a)

(2b)

(2c)

The intramolecular C–H activation reactions involve oxidative addition of an alkylphosphine C–H bond to the model iridium catalyst. Methylphosphino (eq. 2a), ethylphosphino (eq. 2b), and propylphosphino (eq. 2c) are the three ligands studied, giving a three-, four-, and five-membered metallacycle, respectively, as product. Intramolecular C–H activation is very common in the ortho position of aryl phosphines,[8] but activation is also known to occur in aliphatic phosphines. For example, $Fe(PMe_3)_4$ undergoes activation at one of the methyl groups, producing the three-membered metallacycle with a hydride ligand.[9] The terminal CH_3 in a propylphosphino ligand in $Cp*Rh(H)(C_3H_7)(PMe_2Pr)_2$ activates, giving the five-membered metallacycle.[3] Even though the computational models have been kept small for tractability, reactions 2a – 2c are experimentally relevant since it is known they contribute to catalyst degradation.

2. COMPUTATIONAL METHODS

Calculations employ the sequential and parallel versions of the GAMESS program.[10] Effective core potentials (ECPs) and valence basis sets[11] are used for all heavy atoms and the -31G basis set for H. ECPs replace the innermost core orbitals for transition metals (TMs) and all core orbitals for main-group (MG) elements. Thus, nd, $(n+1)s$ and $(n+1)p$ along with ns and np are treated explicitly for the d-block; for the MG ns and np are treated explicitly. Transition metal valence basis sets are quadruple and triple zeta for the sp and d shells, respectively, while MG elements have a double-zeta valence basis. Basis sets for heavy MG elements are augmented with a d-polarization function.

Geometries are optimized using restricted Hartree Fock (RHF) and wavefunctions for closed-shell singlets. Bond lengths and angles for TM complexes are typically predicted to within 1–3 % of experimental models using the present computational scheme, termed RHF/SBK(d), involving complexes in a variety of geometries, formal oxidation states, and metals from the entire transition series.[12] Vibrational frequencies are calculated at stationary points to identify them as minima or transition states.

Møller-Plesset second-order perturbation theory (MP2),[13] at RHF/SBK(d) geometries, is used to calculate the correlation contribution to the total energies. Previous work[4, 5] has shown an [RHF geometry/MP2 energy] scheme to be an efficient route for studying series of related reactions. One should exercise caution in interpreting calculated energies since MP2 tends to overestimate the exothermicity of C–H oxidative addition. Reported energies are corrected for the zero point energy and from absolute zero to 298.15 K (using RHF vibrational frequencies).

The intrinsic reaction coordinate (IRC)[14] is the steepest descent path in mass-weighted Cartesian coordinates from the transition state (TS) to reactants and products. The IRC provides dynamic information about chemical processes and hence interactions that control important transformations. The IRC is a valuable tool in the analysis of catalytic reactions.[4, 5] Calculation of the intrinsic reaction coordinates employed the method of Gonzalez and Schlegel.[15]

Crabtree et al.[16] have analyzed the crystal structures of agostic complexes and performed a Bürgi-Dunitz analysis[17] to construct an experimental C–H activation trajectory. This trajectory is used to follow the C–H distance and M–H–C angle throughout C–H oxidative addition.[16] In a previous computational study of methane and ethane activation, good correspondence was shown between experimental and calculated C–H activation trajectories.[4, 5] The Crabtree trajectory was constructed from agostic complexes and represents a trajectory for intramolecular C–H activation.[16] Thus, it is of interest to compare this trajectory with oxidative addition of methane[4] and ethane,[5] intermolecular CH activation, and β-H transfer, intramolecular C–H activation.[5]

Crabtree et al.[16] defined r_{bp}, eq. 3, where r (0.28) is "the ratio of the H covalent radius to the standard C–H distance," 1.27 Å is the covalent radius of Ir, and r_{bp} is "the covalent radius of the C-H bonding electrons."

$$r_{bp} = [d_{MH}^2 + r^2 d_{CH}^2 - r(d_{MH}^2 + d_{CH}^2 - d_{MC}^2)]^{1/2} - 1.27 \text{ Å} \qquad (3)$$

As r_{bp} decreases the strength of the agostic M$\cdots$H–C interaction increases. In this research S_{total} has been employed (eq. 4) as the reaction coordinate, $x_i(0)$ is the x coordinate at the TS, $x_i(s)$ is the x coordinate at some point along the IRC and so forth.

$$S_{total} = S[(x_i(0) - x_i(s))^2 + (y_i(0) - y_i(s))^2 + (z_i(0) - z_i(s))^2]^{1/2} \qquad (4)$$

Previous work on alkane activation shows that S_{total} and r_{bp} yield nearly identical trajectories.[4]

3. RESULTS AND DISCUSSION

Reactant Complexes

All complexes discussed were optimized under C_1 symmetry. The geometry and bond lengths are nearly identical for the methyl-, ethyl-, and propyl-phosphino reactant complexes (1a, 1b, 1c), and so will be discussed together. The geometry of the reactant complexes is T-shaped. The different substituents on the phosphines did not affect the bond lengths. The hydride bond length is 1.58 Å, consistent with Ir hydrides characterized by neutron diffraction[18, 19, 20] (IrH bond lengths ≈ 1.6 Å). Experimentally, three-coordinate Ir-bis(phosphine) complexes are not isolable. An analysis of a structural database for four-coordinate, trans-Ir^I-bis(phosphine) complexes (monodentate phosphines only) yields a range of Ir-P bond lengths from 2.28–2.35 Å, with an average of 2.32 Å for 22 sample structures.[20] The reactants (**1a–1c**) are calculated to have Ir–P = 2.31 Å. The P–Ir–P angle is the same for the reactants, 172°. The alkyl groups are pointing away from the hydride to minimize steric interactions. In reducing the steric interactions, the phosphine substituents are pointing towards an open coordination site.

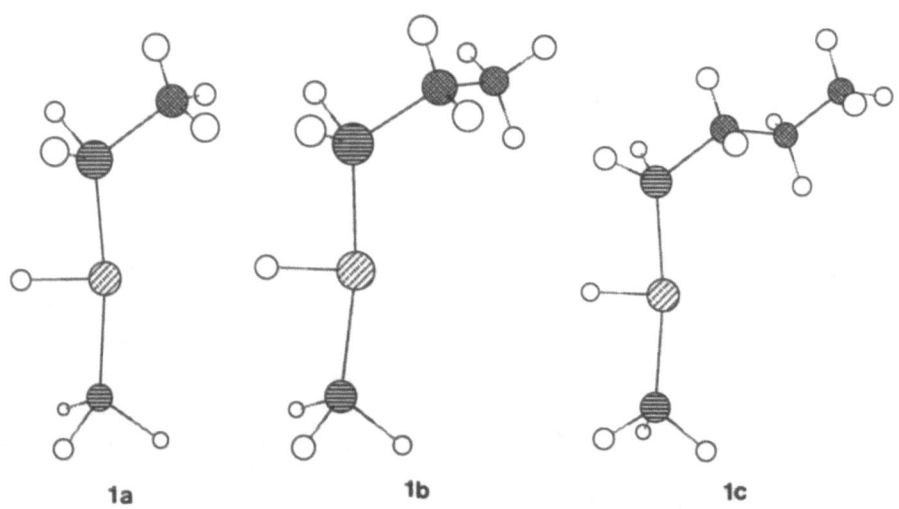

| 1a | 1b | 1c |

Agostic Complexes

No agostic complexes were located for the methyl- and ethyl-phosphino complexes. In both cases the IRCs for oxidative addition led to non-agostic reactants and products. This is likely due to the extra strain on the Ir–P–C angle not being compensated by the agostic interaction. The IRC for the propyl-phosphino complex did not lead to **1c**, but to an agostic complex, **2**. The hydride (1.59 Å) and Ir-P bond lengths (2.31 Å) are unchanged for agostic versus non-agostic isomers. The Ir–H bond length (2.28 Å) is comparable to the previously calculated agostic ethyl complex $(Ir(H)_2(PH_3)_2(C_2H_5)$, Ir–H = 2.23 Å).[5] The CH bond is lengthened slightly in the agostic complex, 1.09 Å in **1c**, and 1.11 Å in agostic **2**. The quantity defined by Crabtree,[16] r_{bp}, was calculated to be 1.24 Å, indicating a weak agostic interaction. This is weaker than the previously calculated agostic interaction and σ-complexes, i.e., the methane σ-complex $Ir(H)(PH_3)_2 \cdots CH_4$ (r_{bp} = 1.15 Å),[4] the ethane σ-complex, $Ir(H)(PH_3)_2 \cdots C_2H_6$ (r_{bp} = 1.13 Å), and the agostic ethyl complex $Ir(H)_2(PH_3)_2(C_2H_5)$ (r_{bp} = 1.02 Å).[5] The propyl arm may not be long enough or flexible enough to fully compensate for the adoption of a favorable geometry for agostic interaction. At the MP2 level agostic **2** is 4 kcal mol^{-1} lower in energy than non-agostic **1c**, considerably less than the binding of ethane to $Ir(H)(PH_3)_2$ in their σ-complex, $\Delta H_{add}(Ir(H)(PH_3)_2 \cdots C_2H_6)$ = −8kcal mol^{-1} (MP2 level).

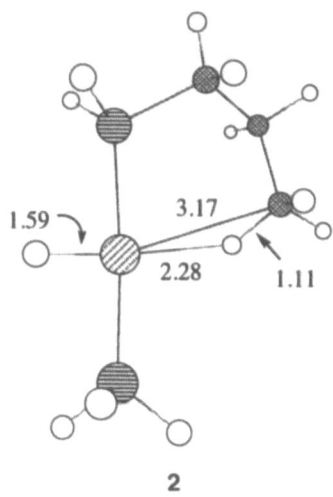

2

Product Complexes

The products for the cyclometallation of the methyl-, ethyl-, and propyl-phosphino complexes are three-, four-, and five-membered metallacycles, respectively, **3**. The cis hydride is constant at 1.55 Å for all three complexes. The difference in ring strain between three- (**3a**) and five-membered (**3c**) rings is obvious. The P–Ir–P angle increases from 155° (**3a**) to 168° (**3b**) to 172° (**3c**). The P–Ir–P angle in **3c** is the same as the reactant, **1c**, indicating no strain in the former. The Ir–C bond length decreases from 2.24 Å (**3a**) to 2.21 Å (**3b**) to 2.20 Å (**3c**). Since the Ir–C bond length decreases, there is an increase in the trans influence on the trans hydride (**3**). The Ir-trans hydride bond length increases from 1.65 Å (**3a**) to 1.67 Å (**3b**) to 1.68 Å (**3c**).

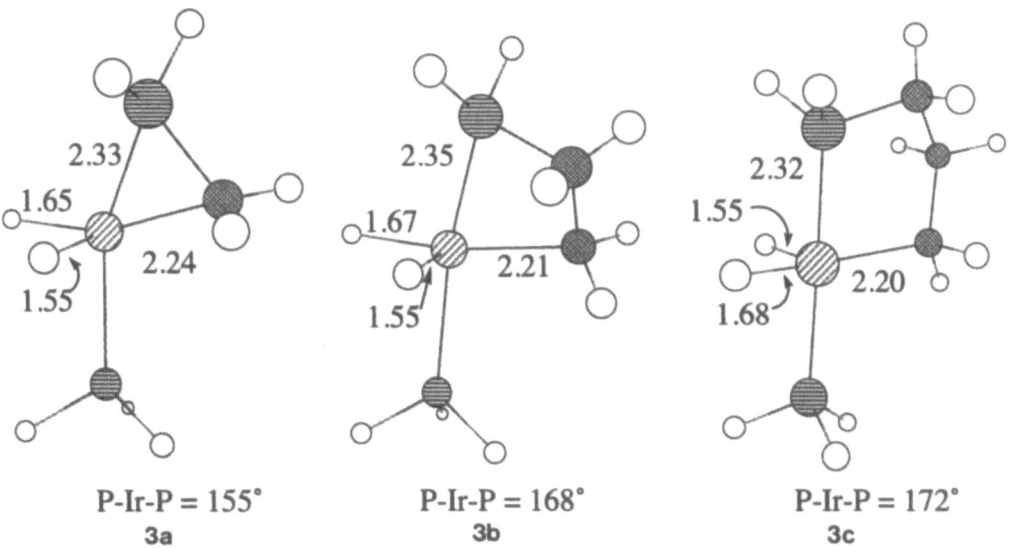

Several Ir(III) cyclometallated complexes have been characterized by X-ray diffraction. Two are three-membered IrPC rings: $IrCl_2(\eta^2\text{-}CH_2PMePh)(PMe_2Ph)_2$[21] (Ir–P = 2.276(6) Å, Ir–C = 2.19(2) Å, P–C = 1.90(3) Å), and $fac\text{-}Ir(\eta^2\text{-}CH_2PPh_2)H[N(SiMe_2CH_2PPh_2)_2]$[22] (Ir–P = 2.241(2) Å, Ir–C = 2.203(7) Å, P–C = 1.808(9) Å). Calculated values for three-membered metallacycle **3a** are Ir–P = 2.33 Å, Ir–C = 2.24 Å, and P–C = 1.80 Å.

The largest difference is < 6 % (most are < 3 %), which is reasonable given differences in ligands and geometry. A five-membered ring has been characterized, $IrCl(H)(\eta^2\text{-}CH_2CMe_2CH_2P^tBu_2)(^tBu_2PCH_2CMe_3)$[23] with Ir–P = 2.402(9) Å, Ir–C = 2.17(2) Å, and P–Ir–P = 167.8(3)°. Calculated values for **3c** are Ir–P = 2.32 Å, Ir–C = 2.20 Å, and P–Ir–P = 172°, in excellent agreement with experiment.

Transition States

The TSs involve intramolecular oxidative addition, and therefore are three-centered (**4a**, **4b**, **4c**), as seen for the β-H transfer TS involving $Ir(H)_2(PH_3)_2(C_2H_5)$.[5] The IrH bond is 1.63 Å in the methyl-phosphino TS (**4a**) and 1.61 Å in both the ethyl- and propyl-phosphino TSs (**4b** and **4c**, respectively). The CH bond decreases with increasing ring size, 1.64 Å (**4a**) to 1.58 Å (**4b**) to 1.54 Å (**4c**). The IrC bond increases from 2.21 Å in **4a** to 2.26 Å in **4b** and **4c**. The IrC and IrH bonds are nearly formed in the TSs, supporting Crabtree's inferred late TS.[6]

The ring strain is not as obvious in the TSs as in the products, due to the P–Ir–P angle bending to accommodate ring formation. The P–Ir–P angle in **4a** is 163°, and increases to 175° in **4b** and 177° in **4c**. The IrH bond is 1.66 Å in **4a**, and 1.68 Å in **4b** and **4c**. Differences in IrC bond strength in the TSs (from ring strain) would be obvious in the hydride bond length due to the trans influence, but IrH bonds change by only 0.02 Å from **4a** to **4c**.

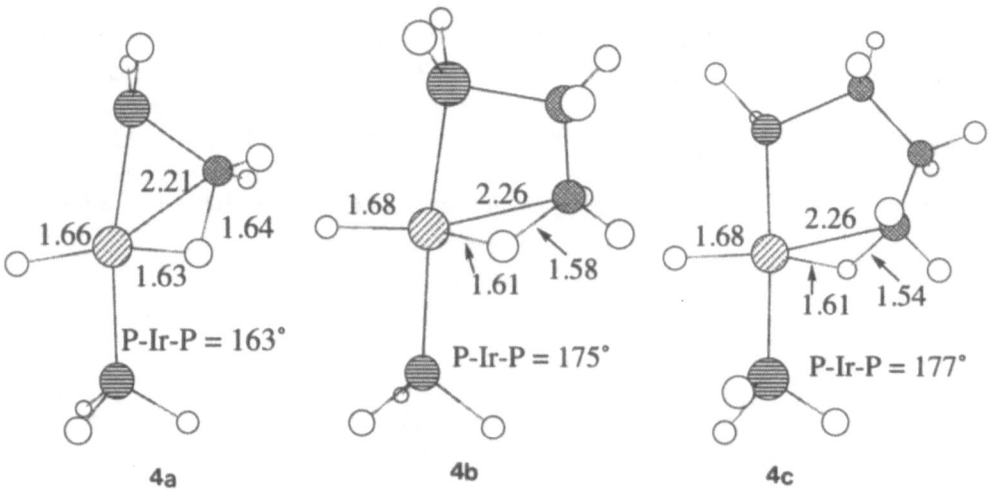

Energetics

The enthalpy of cyclometallation is very dependent on metal and ligand environment. In the case of $Fe(PMe_3)_4$, although a strained three-membered ring is formed, the reaction is spontaneous, with the reactant only being observed in the gas phase by mass spectrometry.[9] The effect of ring strain is very obvious in the energetics of the cyclometallation reaction, **5** (and listed in Table 1). The activation enthalpy (MP2 level of theory) of the methyl-phosphino reaction (**1a** → **4a** → **3a**) is 17 kcal mol^{-1}, 9 kcal mol^{-1} for the ethyl-phosphino reaction (**1b** → **4b** → **3b**), and −7 kcal mol^{-1} for the propyl-phosphino reaction (**1c** → **2** → **4c** → **3c**), Figure 1. The methyl- and ethyl-phosphino reactions are endothermic by 7 kcal mol^{-1} and 2 kcal mol^{-1}, respectively, while the propyl-phosphino reaction is exothermic by 14 kcal mol^{-1}.

The propyl-phosphino cyclometallation is intramolecular CH activation with an agostic interaction, and therefore is of interest to compare with the previously studied β-H transfer.[5] In the former, the reaction is exothermic by 10 kcal mol^{-1} starting with agostic complex **2**, while

Table 1. RHF and MP2 energies of stationary points

Complex, Figure #	RHF[a] (Hartrees)	MP2[a] (Hartrees)	H[b] (kcal mol^{-1})
Ir(PH$_3$)(PH$_2$CH$_3$)(H), **1a**	-127.3645	-127.8620	67.638
[Ir(PH$_3$)(PH$_2$CH$_2$···H···)(H)]‡, **4a**	-127.2910	-127.8299	63.998
Ir(PH$_3$)(PH$_2$CH$_2$-)(H)$_2$, **3a**	-127.3262	-127.8458	64.797
Ir(PH$_3$)(PH$_2$C$_2$H$_5$)(H), **1b**	-134.0121	-134.6317	87.628
[Ir(PH$_3$)(PH$_2$C$_2$H$_4$···H···)(H)]‡, **4b**	-133.9523	-134.6120	83.952
Ir(PH$_3$)(PH$_2$C$_2$H$_4$)(H)$_2$, **3b**	-133.9845	-134.6241	84.507
Ir(PH$_3$)(PH$_2$C$_3$H$_7$)(H), **1c**	-140.6619	-141.4029	107.428
Ir(PH$_3$)(PH$_2$C$_3$H$_6$···H···)(H), **2**	-140.6596	-141.4086	107.310
[Ir(PH$_3$)(PH$_2$C$_3$H$_6$···H···)(H)]‡, **4c**	-140.6232	-141.4082	104.073
Ir(PH$_3$)(PH$_2$C$_3$H$_6$)(H)$_2$, **3c**	-140.6555	-141.4208	104.557

[a] 1 Hartree = 627.510 kcal mol^{-1}
[b] Correction for zero point energy and from absolute zero to 298.15 K calculated from RHF vibrational frequencies.

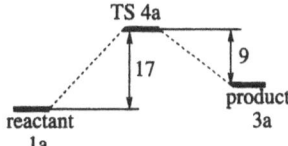

a. methyl-phosphino reaction

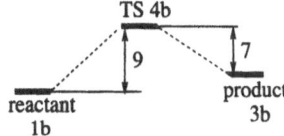

b. ethyl-phosphino reaction

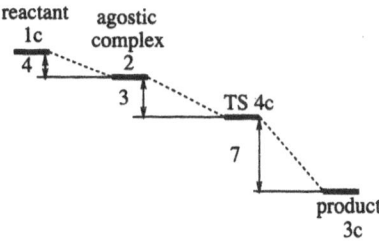

c. Propyl-phosphino reaction

Figure 1. Calculated reaction enthalpies (in kcal mol^{-1}) for cyclometallation.

77

the β-H transfer reaction is exothermic by 16 kcal mol^{-1}. The product of the β-H transfer is an olefin complex. Based on the CC bond length (1.38 Å), the olefin complex is more a π-complex than a metallacyclopropane,[5] so it does not require the energy needed to form a small ring system.

Intrinsic Reaction Coordinate

The IRC was calculated for each cyclometallation reaction, and Crabtree trajectories[16] were constructed (Figure 2 – 4). The methyl-phosphino IRC (Figure 2) and the ethyl-phosphino IRC (Figure 3) are very similar. The CH bond "break point" ($S_{total} \approx -1$ bohr amu$^{1/2}$) is very close to the transition state ($S_{total} = 0$ bohr amu$^{1/2}$) for these two reactions. The IrC and IrH bonds are nearly formed at the TS, and the CH bond stays constant until the transition state. For the ethyl-phosphino complex, the Ir–H–C angle also remains constant until close to the transition state. The CH$_3$ group in the ethyl-phosphino complex, although not an agostic complex, is in a good orientation for CH activation in the reactant. The CH$_3$ group in the methyl-phosphino complex is pointed away from the metal. Much of the initial energy is used to decrease the Ir–P–C angle, moving the CH$_3$ group into the appropriate orientation for CH activation. This is evident in the large increase in Ir-H-C angle approaching the transition state, and accounts for the large activation energy.

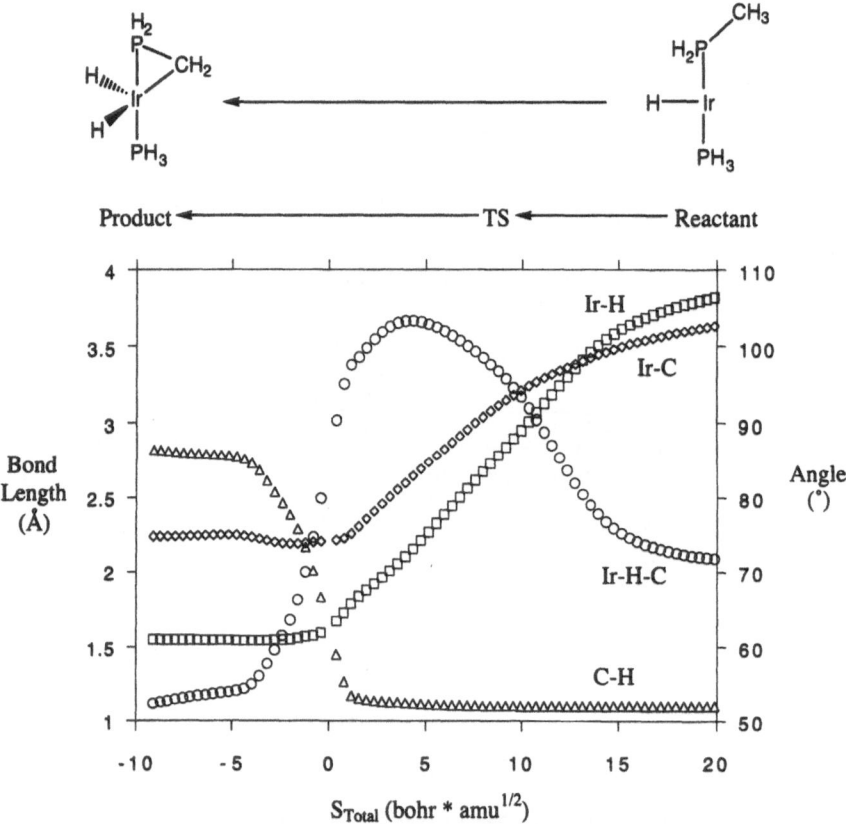

Figure 2. Trajectory for cyclometallation of the methyl-phosphino complex.

The Crabtree trajectory for the propyl-phosphino cyclometallation reaction (Figure 4) is roughly the same as the trajectory for the methyl- and ethyl-phosphino reactions,[4,5] even

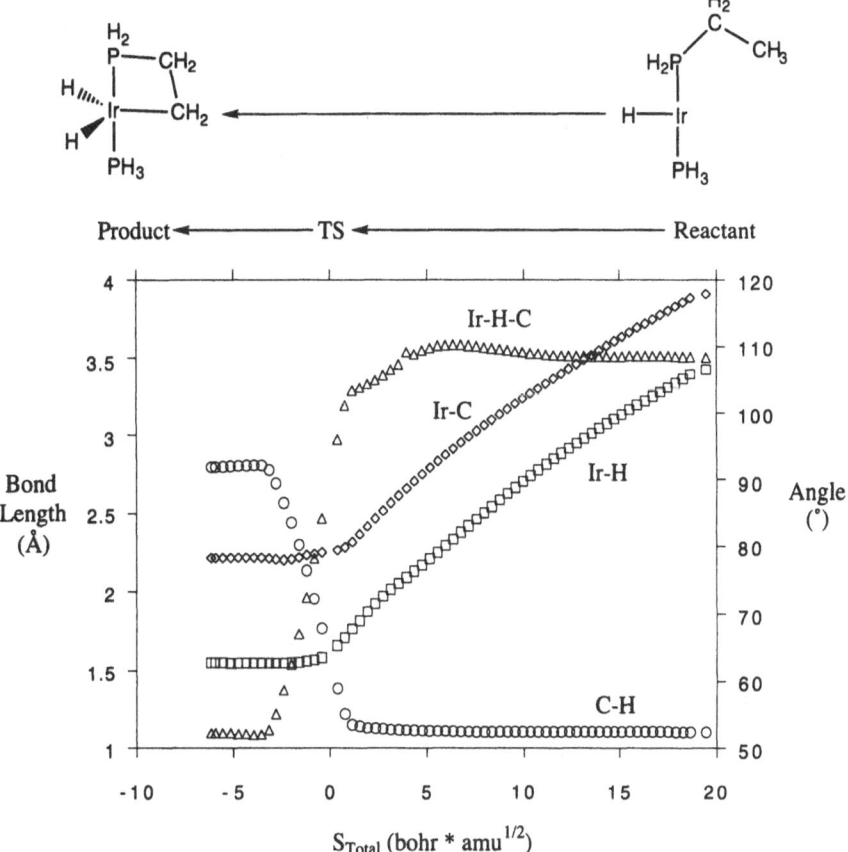

Figure 3. Trajectory for cyclometallation of the ethyl-phosphino complex.

though the CH break point is around $S_{total} = -5$ bohr amu$^{1/2}$. Undue significance should not be ascribed to the position of various nonstationary points on the IRC. The transition state is still late, as in the other oxidative addition reactions.[4,5] The qualitative information contained in the trajectory is identical to the other trajectories. The CH bond stays constant until very close to the break point, IrC and IrH decrease until $S_{total} \approx -4$ bohr amu$^{1/2}$, and remain constant after the breaking point, and the CH bond and Ir–H–C angle sharply increase and decrease, respectively, at the breaking point. Of greater significance is that the behavior of the Ir–H–C angle in cyclometallation of the propylphosphino complex is more akin to that seen in oxidative addition of ethane. This suggests that for the alkylphosphine with a three- carbon chain there is sufficient conformational flexibility so that the dynamics of *intramolecular* C–H activation step are similar to those for analogous *intermolecular* processes. Presumably, this conclusion also applies to alkylphosphines with carbon chains longer than three.

4. CONCLUSIONS

Cyclometallation, a potential pathway for catalyst degradation, has been studied computationally for the model catalyst Ir(H)(PH$_3$)(PH$_2$R), where R = methyl, ethyl, or propyl. Several conclusions have been reached about cyclometallation reactions.

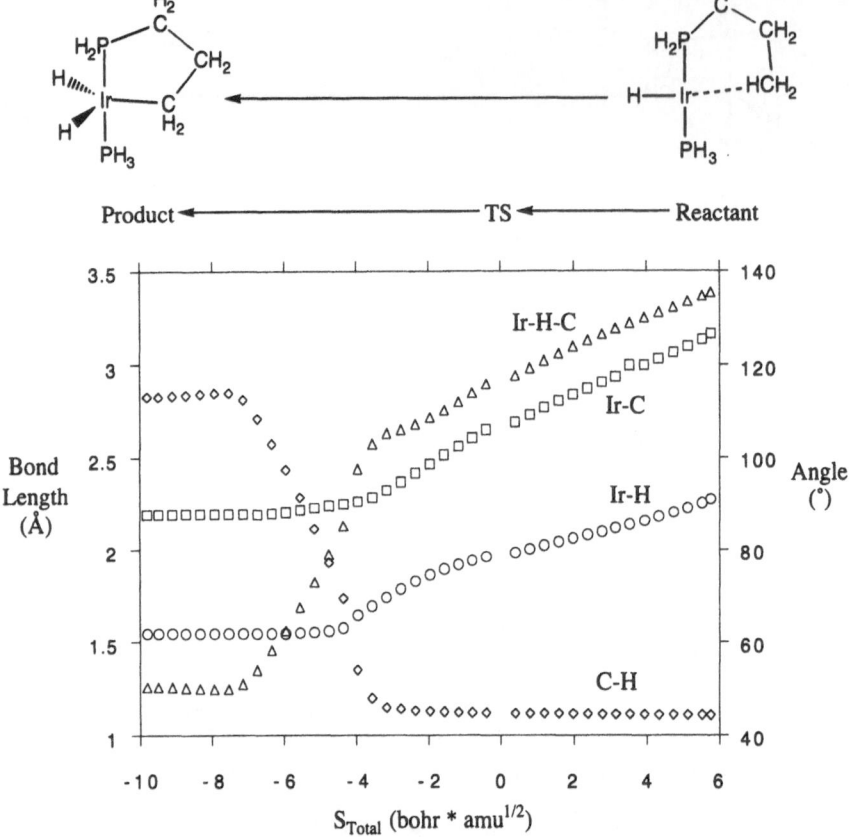

Figure 4. Trajectory for cyclometallation of the propyl-phosphino complex.

1) When the phosphine substituent is methyl or ethyl, an agostic complex prior to CH activation is not formed. Although an interaction between the CH bond and the metal is needed prior to CH activation, an agostic complex is not a minimum. The strain that would be present in an agostic complex is not compensated for by the agostic interaction between Ir and the C–H bond of the methyl or ethyl substituent; also, the IRC did not lead to an agostic complex, but to non-agostic reactants. The IRC for the propyl-phosphino complex did lead to an agostic complex, which is weaker than intermolecular and intermolecular analogues.

2) Much of the initial energy expended for cyclometallation of the methyl-phosphino, and to some extent the ethyl-phosphino, is required to move the C–H bond to be activated into the appropriate orientation. The reactant for the propyl-phosphino cyclometallation is an agostic complex, which therefore has the terminal CH_3 group in the required orientation.

3) The methyl- and ethyl-phosphino cyclometallations are endothermic with positive activation barriers, while the propyl-phosphino reaction is exothermic with a negative activation barrier. The size of the ring in the product affects the cyclometallation activation barrier. The calculations suggest the resistance of the methyl- and ethyl-phosphino complexes to cyclometallation is due in part to the lack of an agostic precursor. For the shortest carbon chains, the potential agostic interaction is more than offset by increased strain from reducing the P–Ir–P angle below 180°. Although examples are known where three- and

four-membered rings are formed (i.e., $Fe(H)(PMe_3)_3(\eta^2\text{-}CH_2PMe_2)^9$), the above data indicates smaller alkyl substituents on the phosphines are a means to reduce catalyst degradation from cyclometallation.

5. ACKNOWLEDGEMENTS

The authors acknowledge the support of this research by the United States National Science Foundation through grants CHE-9314732 and CHE-9614346. M.T.B. acknowledges the van Vleet Memorial Foundation for fellowship support.

REFERENCES

1. C. L. Hill, ed. "Activation and Functionalization of Alkanes," Wiley, New York (1989).
2. For example, see F. A. Cotton and G. Wilkinson. "Advanced Inorganic Chemistry," 5th Ed., Wiley, New York (1988), p. 432–439.
3. W. D. Jones and F. J. Feher, *J. Am. Chem. Soc.* 107:620 (1985).
4. T. R. Cundari, *J. Am. Chem. Soc.* 116:340 (1994).
5. M. T. Benson and T. R. Cundari, *Inorg. Chim. Acta* 259:91 (1997).
6. R. H. Crabtree, in reference 1, p. 69.
7. M. J. Burk and R. H. Crabtree, *J. Am. Chem. Soc.* 109:8025 (1987).
8. G. W. Parshall, *Acc. Chem. Res.* 3:139 (1970).
9. J. W. Rathke and E. L. Muetterties, *J. Am. Chem. Soc.* 97:3272 (1975).
10. M. W. Schmidt, K. K. Baldridge, J. A. Boatz, J. H. Jensen, S. Koseki, N. Matsunaga, M. S. Gordon, K. A. Nguyen, S. Su, T. L. Windus, S. T. Elbert, J. Montgomery, and M. Dupuis, *J. Comput. Chem.* 14:1347 (1993).
11. M. Krauss, W. J. Stevens, H. Basch, and P. G. Jasien, *Can. J. Chem.* 70:612 (1992).
12. T. R. Cundari and M. S. Gordon, *Coord. Chem. Rev.* 147:87 (1996).
13. C. Møller and M. S. Plesset, *Phys. Rev.* 46:618 (1934).
14. D. G. Truhlar and M. S. Gordon, *Science* 249:491 (1990).
15. C. Gonzalez and H. B. Schlegel, *J. Chem. Phys.* 90:2154 (1989).
16. R. H. Crabtree, E. M. Holt, M. Lavin, and S. M. Morehouse, *Inorg. Chem.* 24:1986(1985).
17. H. B. Bürgi and J. D. Dunitz, *Acc. Chem. Res.* 16:153 (1983).
18. R. Bau and G. Teller, *Struct. Bonding (Berlin)* 44:1 (1981).
19. L. Garlaschelli, S. T. Kahn, R. Bau, G. Longoni, and T. F. Koetzle, *J. Am. Chem. Soc.* 107:7212 (1985).
20. Structures were taken from the Cambridge Structural Database (version 5.1.3, F. H. Allen and O. Kennard, *Chem. Des. Automation News* 8:31 (1993).). A search was done for four-coordinate, Ir(I) complexes with two monodentate phosphine ligands (trans orientation). The search was refined to include only complexes for which atomic coordinates were available, and excluded experimental structures with disorder, R factors $\geq 10\%$, or which were polymeric.
21. M. D. Fryzuk, K. Joshi, R. K. Chadha, and S. J. Rettig, *J. Am. Chem. Soc.* 113:8724 (1991).
22. S. Al-Jibori, C. Crocker, W. S. McDonald, and B. L. Shaw, *J. Chem. Soc., Dalton Trans.* (1981) 1572.
23. L. Dahlenburg and N. Höck, *Inorg. Chim. Acta* 104:L29 (1985).

C–H···O HYDROGEN BONDS IN ORGANOMETALLIC CRYSTALS

Dario Braga and Fabrizia Grepioni

Dipartimento di Chimica G. Ciamician
Università di Bologna
Via Selmi 2
40126 Bologna, Italy

1. INTRODUCTION

The experimental[1] and theoretical[2] generation of crystal architectures is attracting the interest of an increasing number of research groups. The task is that of *making crystals with a purpose*. This implies design, synthesis, characterization and utilization of crystalline materials with predefined assembly of molecules and ions that result in useful collective crystalline properties (magnetism,[3] conductivity and superconductivity,[4] charge transfer,[5] NLO applications[6] etc.).

No matter whether the approach is practical or theoretical, a successful crystal engineering strategy implies the capacity of controlling the forces acting between molecules or ions.[7] In molecular crystals, these forces involve mainly peripheral atoms at contact distance from neighboring atoms in the crystals. For practical purposes the peripheral interactions are usually categorized in three main classes, namely van der Waals, Coulombic and hydrogen bonding interactions. This commonly accepted subdivision, though serving well the necessity of describing and discussing crystal engineering models, is sometimes dangerously misleading. It is, in fact, easy to forget the basic thermodynamics behind the stability of a given crystalline material. On a macroscopic scale, the free energy minimum (not necessarily the global minimum though) corresponding to a given crystal structure is the result of an overall intermolecular bonding optimization, viz. any partitioning of the global intermolecular interaction in (usually additive) contributions is artificial and must be done with great caution.

Albeit important, strength is not the only prerequisite of a "useful" intermolecular bond. Directionality, i.e. the ability of organizing in space the direction of action of the bond, is crucial in crystal engineering exercises. A strong intermolecular bond with little vectorial characteristics can be less useful than a weak bond acting only within a well defined topology. This is the underlying reason for the primary role played by hydrogen bonds in crystal engineering.[8] The hydrogen bond combines strength and directionality, hence it is the interaction of choice in most crystal building strategies.[9]

2. HYDROGEN BONDS IN ORGANOMETALLIC CRYSTALS

The strength of the hydrogen bond depends on the nature of the donor and acceptor atoms. The energy of this bond, generally described as a three-center four-electron interaction, is dominated by electrostatic factors. Several useful compilations of hydrogen bonds ranked in order of strength and/or length are available in the literature.[10] There is no clear demarcation, however, between strong and weak bonds and we shall not attempt to pursue this matter in the context of this article. In most general terms, and with the caveat raised above, homoatomic interactions (such as $O-H\cdots O$, $N-H\cdots N$) tend to be stronger than heteroatomic hydrogen bonds such as $N-H\cdots O$, $C-H\cdots O$, $O-H\cdots N$ etc. The effect of charge needs also to be taken into account as it has a great influence on the strength of these interactions because of their electrostatic nature. Negatively charged,[11] positively charged,[12] as well as resonant[13] hydrogen bonds have been identified. More recently, the role of charge assisted bonds $(X-H^{\delta+}\cdots Y^{\delta-})$ between donor and acceptor groups belonging to ions carrying opposite charges has emerged as a key structural feature in ionic crystals.[14]

The role of the carbon atom is somewhat controversial, and the very existence of proper hydrogen bonds involving this atom as a donor has been matter of debate for a long time. The matter appears now definitely settled as more and more evidence of their bonding nature have been accumulated during the past few years. This is witnessed by the large number of publications devoted to $C-H\cdots O$ interactions.[15] Both spectroscopic[16] and diffraction data provide evidence for the following (rough) order of decreasing acidity: $C(sp)-H > C(sp^2)-H > C(sp^3)-H$. This acidity can be tuned by the presence of electron withdrawing groups. In terms of accepting capacity, electron rich systems as the triple bond in alkynes,[17] or π-delocalized systems of benzene, phenyl groups and other aromatic carbocycles can accept hydrogen bonds from suitable donors.[18]

These bonds are, in general, weaker than conventional $O-H\cdots O$, $N-H\cdots N$, or $O-H\cdots N$ etc. Theoretical estimate of the strength of the $C-H\cdots O$ bond by means of various empirical,[19] semi-empirical[20] and *ab-initio* methods[21] indicates a range of 5–10 kJ mol^{-1} for interactions of the $C-H\cdots O$ type. However, it is important to keep in mind that most hydrogen bonded molecules usually carry a limited number of hydrogen bond donor/acceptor systems (e.g. dicarboxylic acids, aminoacids, amides, diols etc.), whereas the number of potential donor systems ($C_{sp}-H$, $C_{sp^2}-H$ and $C_{sp^3}-H$) is generally much larger. If enough acceptor sites are available, the cooperative effect of a large number of weak bonds may equal that of one or two strong hydrogen bonds.

This cooperation is by far more important in organometallic chemistry.[22] Inorganic ligands such as CO (but also CN^-, NO^+, CS etc.) as well as organic systems such as arenes and cyclopentadienyl ligands are typical ancillary ligands in organometallic and coordination chemistry. They provide *a large number* of donor (the $C_{sp^2}-H$ system) and acceptor (the η^1, μ_2 or μ_3-CO ligands) groups to form a profusion of hydrogen bonding interactions of the $C-H\cdots O$ type. Although these interactions are individually weak, their collective effect can sum up to that of strong hydrogen bonds.

This contribution will focus on $C-H\cdots O$ bonds in organometallic crystals through a number of examples coming from our own work or extracted from the Cambridge Structural Database.[23] This work is part of our systematic investigation of the way neutral and charged organometallic molecules and clusters self-recognize and self-assemble in the solid state.[24] The objective has been that of achieving sufficient basic knowledge to be able to *choose* intermolecular interactions to engineer novel crystalline materials on the basis of shape, size, and structural functionality.[25]

As a part of our study of hydrogen bonding in organometallic crystals we have shown, in collaboration with Gautam Desiraju, that organometallic complexes and clusters afford both

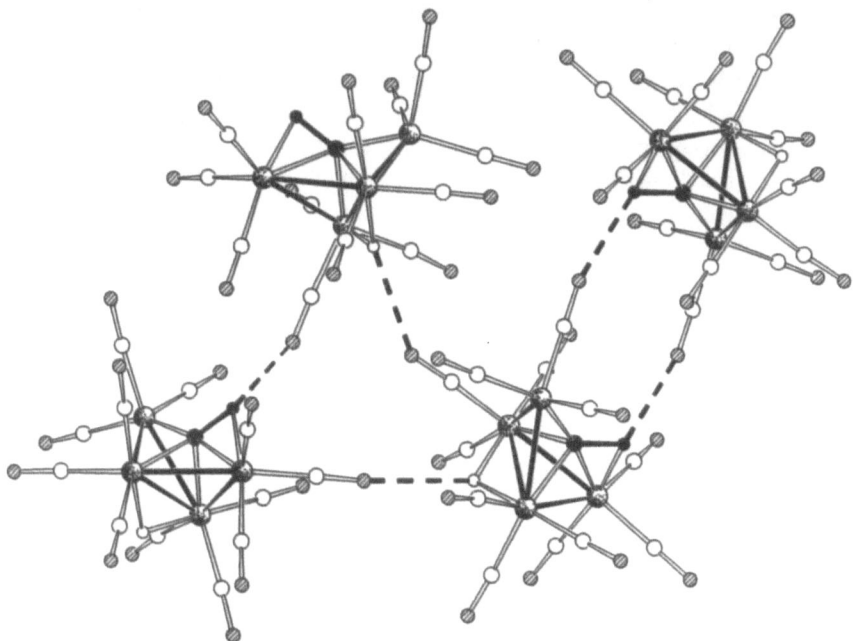

Figure 1. C−H···O interactions involving the agostic C−H system and the hydride ligand in crystalline in HFe$_4$(η^2−CH)(CO)$_{11}$(PPh$_3$).

"new" hydrogen bonding acceptors and "new" hydrogen bonding donors.[26] One of the best examples is the CO-ligand which behaves as a hydrogen bonding acceptor of *tunable* basicity thanks to the possibility of varying the coordination mode of the ligand with the metal centers. We have shown, both on the basis of CSD analysis and by examining families of closely related complexes, that the CO ligand participates in C−H···OC interactions whose strength follows roughly the order μ_3−CO > μ_2−CO > η^1−CO, which corresponds to the order of decreasing basicity of the ligand. Metal bound hydrogen atoms, on the other hand, possess an amphoteric behavior: μ_3−H and μ_2−H ligands, for example, can act as hydrogen bonding donors in M−H···OC interactions,[27] whereas terminal M−H systems may behave, depending on the co-ligands on the complexes, as weak hydrogen bonding acceptor sites with suitable donor groups.

3. ORGANOMETALLIC M−(σ)−C−H DONORS

Organometallic molecules are those characterized by the presence of a direct M−C bond between a transition metal atom (M) and carbon.[28] If the carbon atom is bound only to metal atoms, the word "carbide" is usually employed.[28] Carbide complexes may be of interstitial or exposed nature depending on whether the C-atom is completely encapsulated within a metal cage (a transition metal cluster) or bound on the periphery. In this latter case, the C-atom is known to exhibit a strong nucleophilic behavior, which is reflected in a high acidity of the C−H system when the carbide atom is protonated.

The butterfly methylidyne clusters HFe$_4$(η^2−CH)(CO)$_{12}$ and HFe$_4$(η^2−CH)(CO)$_{11}$ (PPh$_3$), characterized by the presence of an agostic Fe···(H−C) interaction, provide a useful example of the direct participation of the C−H system in intermolecular C−H···O bonds.[29] The C−H···O bond in crystalline HFe$_4$(η^2−CH)(CO)$_{12}$ was first discussed by Muetterties *et al.*[29b] The bonds formed by the two crystallographically independent molecules involve the terminal CO-ligand and the methylidyne H-atom (see Figure 1) and have slightly different lengths [2.58 and 2.48 Å, respectively]. In $HFe_4(\eta^2-CH)(CO)_{11}(PPh_3)$, $C-H\cdots O$ interactions are not possible because the methylidyne H-atom is completely embedded within the ligand shell and screened from the surroundings by one phenyl group. Both systems, on the other hand, show the presence of intermolecular $(Fe)-H\cdots OC$ interactions between the H(hydride) atoms and the CO-ligands. The discussion of hydrogen bonds involving metal centers, though relevant in the context of organometallic crystal engineering, is beyond the scopes of this article. The reader is addressed to refs. 27 and 29.

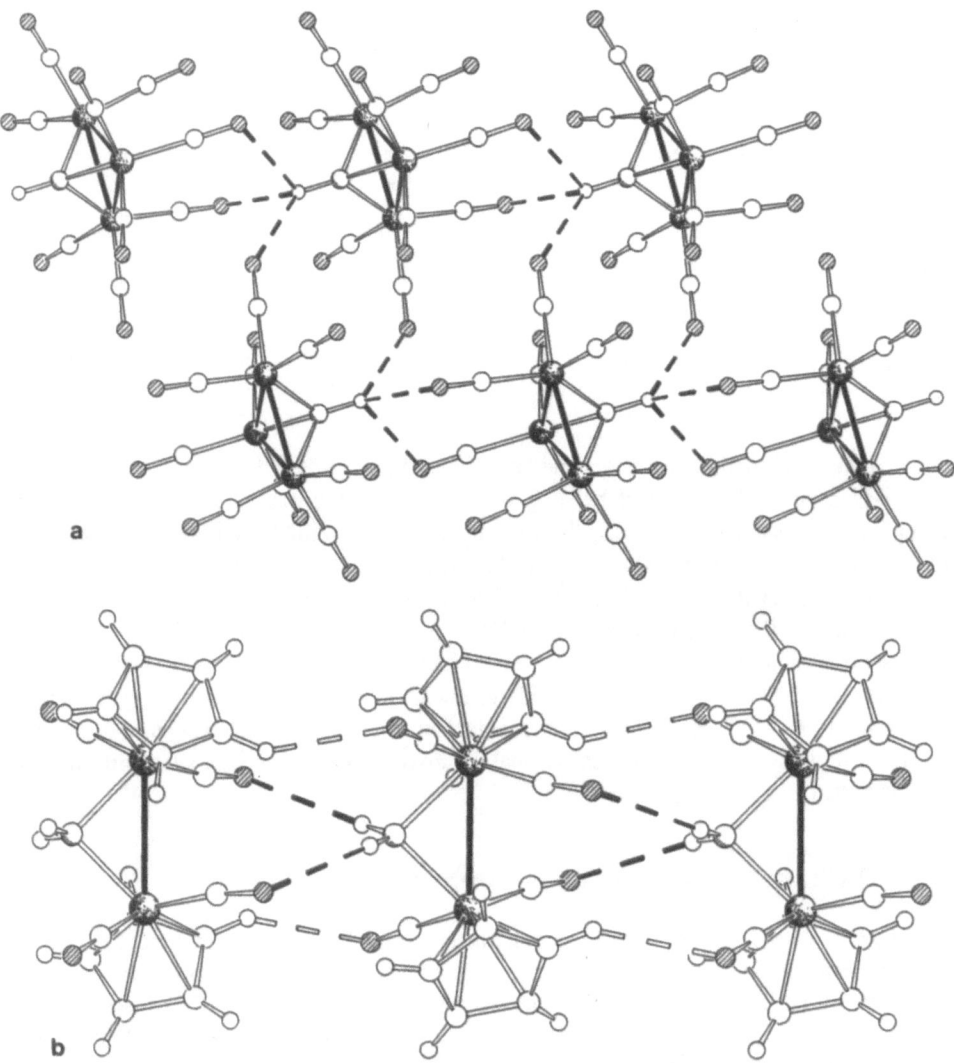

Figure 2. $C-H\cdots O$ intermolecular bonds involving the (a) methylidyne hydrogen atom in crystalline $[Co_3(\mu_3-CH)(CO)_9]$ and (b) the methylene and cyclopentadienyl hydrogen atoms in crystalline $[Cp_2Mn_2(CO)_4(\mu_2-CH_2)]$.

Several methylidyne and methylene clusters have been synthesized and structurally characterized and the bonding of μ_3-CH and μ_2-CH_2 ligands with transition metal clusters has been extensively studied both experimentally[30] and theoretically.[31] Extended Hückel and Fenske–Hall theoretical analyses describe methylidyne and methylene ligands as sp and sp^2 carbon hybrids, respectively.

In order to explore the polarization of the C−H bond or, which is the same, the acidity of the carbon bound H-atom via its involvement in hydrogen bonds in the solid state the CSD has been searched for evidence of the participation of the methylidyne and methylene ligands in hydrogen bonding interactions.[32] The hydrogen bonding in crystalline $[Co_3(\mu_3-CH)(CO)_9]$[33] and $[Cp_2Mn_2(CO)_4(\mu_2-CH_2)]$[34] will be used as representative examples.

The trinuclear methylidyne cluster $[Co_3(\mu_3-CH)(CO)_9]$[33] possesses only terminal CO-ligands and the methylidyne ligand symmetrically spanning the triangular cluster. The μ_3-CH ligand participates in a trifurcated intermolecular $\mu_3-CH\cdots O$ interaction with the CO-ligands $[(C)H\cdots O$ distances in the range 2.50–2.63 Å] belonging to two neighboring molecules as shown in Figure 2a. Polyfurcation is often observed when there is great unbalance between number of donors and number of acceptors as is the case with $[Co_3(\mu_3-CH)(CO)_9]$. The methylidyne acidity matches the basicity of the CO ligand which is almost invariably present as co-ligand in transition metal cluster complexes of this type. Analogously, the μ_2-CH_2 ligand in $[Cp_2Mn_2(CO)_4(\mu_2-CH_2)]$[34] forms a $\mu_2-CH_2\cdots O$ interaction $[(C)-H\cdots O\ 2.56$ Å] (see Figure 2b). The presence of (Cp)H$\cdots$O bonds of comparable length $[(C)-H\cdots O\ 2.50‘$Å] permits an internal comparison between methylene and cyclopentadienyl ligands. In general $\mu_2-CH_2\cdots O$ and (Cp)H$\cdots$O interactions have similar length while they tend to be slightly longer than μ_3-CH$\cdots$O ones. The two groups are therefore comparable with the alkene and alkyne groups in organic crystals.

4. ORGANOMETALLIC M−(π)−C−H DONOR

Aromatic carbocycles, alkynes and alkenes are very common ligands in organometallic chemistry. The bonding interaction with the metal atom(s) is described with the well known mechanism based on σ-donation/π-back donation of electron density between π and π^* of the ligands and suitably oriented d-orbitals on the metal atom. Unsaturated π-ligands of this type are formed mainly by sp^2 and sp hybridized carbon atoms. As mentioned above, hydrogen atoms bound to unsaturated carbon atoms are more acidic than those bound to tetrahedral carbons[16] and can participate in inter- and intramolecular hydrogen bonding with suitable acceptor groups. In addition there is the possibility that the C−H donor is part of a cation and the donor belongs to a counterion. In these cases the C−H$\cdots$O bonds are reinforced by the difference in charge between ions (see below).

An interesting manifestation of the C−H$\cdots$O bonding can be seen in crystals of the cis-, and trans-isomers of $(\eta^5-C_5H_5)_2Fe_2(CO)_2(\mu-CO)_2$.[35] Both isomers carry two cyclopentadienyl ligands as well as two terminal and two bridging CO ligands. C−H$\cdots$O hydrogen bonds are formed between cyclopentadienyl hydrogens and the carbonyl acceptors. The cis-isomer is in general position in the unit cell hence the different packing environments around the two C_5H_5-rings are reflected in the presence of different patterns of C−H$\cdots$OC interactions, as shown in Figure 3, whereas the site symmetry of the trans-isomer both C_5H_5-ligands have crystallographically identical surroundings and experience the same type of interaction. In the crystal of the trans-isomer each molecule forms C−H$\cdots$O interactions with other four surrounding molecules. This difference accounts for the different librational and reorientational motion of the two C_5H_5-ligands in the solid state. The two rings in the cis-isomer not only have appreciably different mean-square librational amplitude of motion about equilibrium positions (302.8 and 62.4 deg^2) but also undergo reorientational jumping motions with different activation energies (7.2 versus 15.8 kJ mol^{-1})[36a] and different potential energy barriers (7.9 versus 17.6 kJ mol^{-1}).[36b] The different motional freedom can be easily ascribed to the

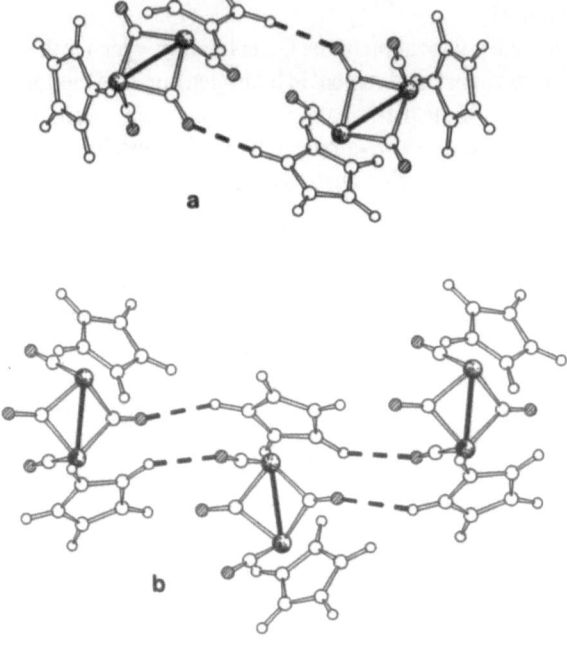

Figure 3. The different patterns of short C−H···O interactions involving the two crystallographically independent C_5H_5-ligands in cis-$(\eta^5-C_5H_5)_2Fe_2(CO)_2(\mu-CO)_2$.

presence of different sets of C−H···O interactions, with shorter (i.e. stronger) bonds with the cyclopentadienyl ligand that moves less. This observation clearly demonstrates the bonding effect that, though weak, is exerted by C−H···O interactions.

Other examples of the effect of C−H···O bonds on the dynamical behavior of atoms in the solid state have been provided by Steiner[37] and by us in collaboration with G. R. Desiraju[38] by investigating organic alkynes and transition metal clusters. Such evidence is important because there is still skepticism as to whether some of these short contacts are actually repulsive in nature and are stabilized by other attractive interactions in the crystal, in other words whether they are the forced consequences of packing rather than packing directors.

Another pair of prototypical organometallic molecules that have been reinvestigated in terms of intermolecular bonding is provided by the chromium complexes $(\eta^5-C_5H_5)_2Cr_2(CO)_6$[39] and $(\eta^5-C_5H_5)_2Cr_2(CO)_4$.[40] Both complexes possess only terminal CO-ligands that accept hydrogen bond formation from the cyclopentadienyl ligands, as shown in Figure 4. The network in $(\eta^5-C_5H_5)_2Cr_2(CO)_6$ is based on centrosymmetric dimers linked by pairs of C−H···O interactions similar to those adopted by −COOH groups.

Arene hydrogens are also invariably involved in C−H···O bonds. We have reported previously several cases in which C−H···O interactions play a detectable role in determining molecular and crystal structure of arene complexes. The relationship between molecular and crystal structures of the benzene carbonyl cluster $Ru_3(CO)_9(\mu_3:\eta^2:\eta^2:\eta^2-C_6H_6)$[42] and the 1,3,5-trithiacyclohexane (TCH, hereafter) complex $Ru_3(CO)_6(\mu-CO)_3(\mu_3-S_3C_3H_6)$[43] has been studied and the presence of bridging CO's in this latter complex has been related to the stabilization provided by C−H···O bonds, as shown in Figure 5.[44]

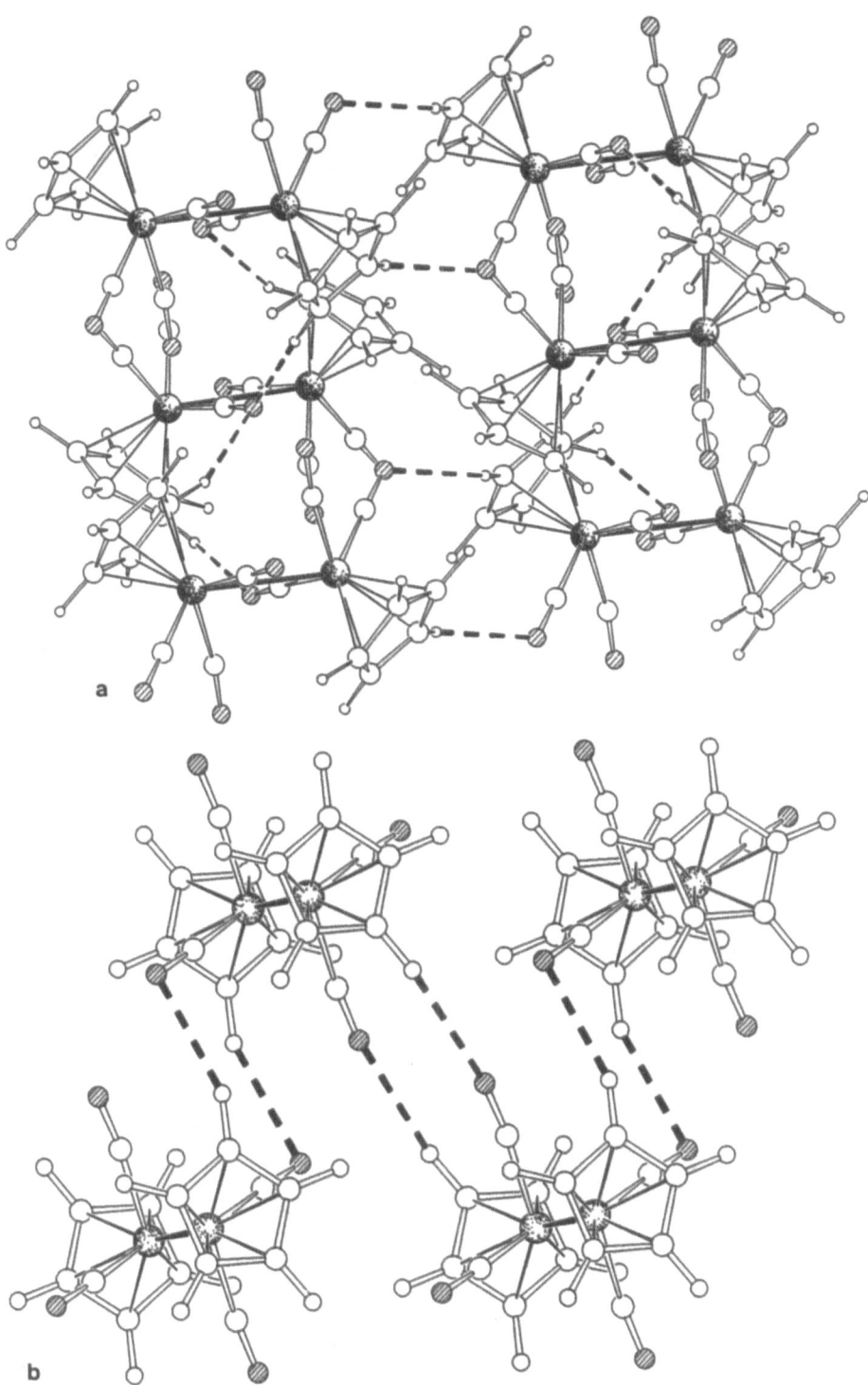

Figure 4. C−H···O hydrogen bond networks in crystalline $(\eta^5-C_5H_5)_2Cr_2(CO)_6$ (a), and $(\eta^5-C_5H_5)_2Cr_2(CO)_4$ (b).

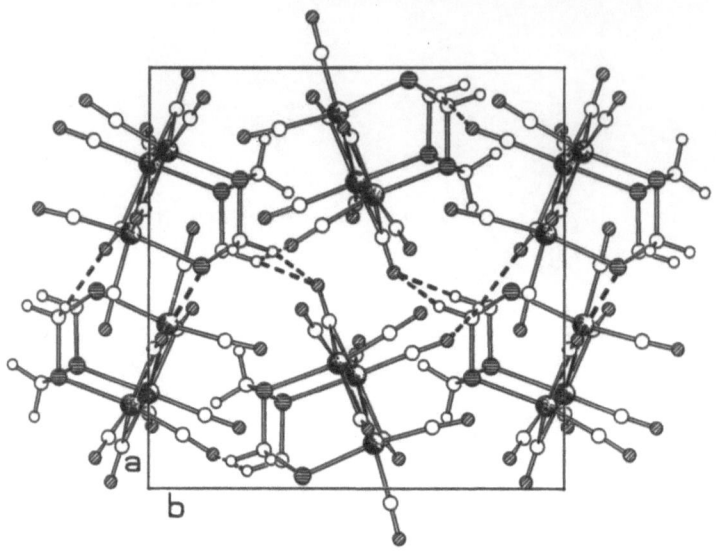

Figure 5. C$-$H$\cdots$O hydrogen bonds in crystalline Ru$_3$(CO)$_6$($\mu-$CO)$_3$(μ_3-S$_3$C$_3$H$_6$).

The TCH ligand has been first shown to participate in hydrogen bonds of the C$-$H$\cdots$O(CO) type in crystals of the two isomers Ir$_4$(CO)$_6$($\mu-$CO)$_3$(μ_3-S$_3$C$_3$H$_6$) and Ir$_4$(CO)$_9$(μ_3-S$_3$C$_3$H$_6$).[45a] In the case of the bridged isomer the bridging CO ligands form shorter C$-$H$\cdots$O interactions than terminal CO's in keeping with the order of basicity (see below). S-atom lone pairs also take part in hydrogen bonding interaction. Other families of TCH clusters have been investigated.[45b]

The size of the metal cluster has no effect on the length of C$-$H$\cdots$O bonds. In the cases of the isomeric pairs Ru$_5$C(CO)$_{12}$(η^6-C$_6$H$_6$) and Ru$_5$C(CO)$_{12}$($\mu_3-\eta^2$:η^2:η^2-C$_6$H$_6$),[46] Ru$_6$C(CO)$_{11}$(η^6-C$_6$H$_6$)$_2$ and Ru$_6$C(CO)$_{11}$(η^6-C$_6$H$_6$)($\mu_3-\eta^2$:η^2:η^2-C$_6$H$_6$)[47] the relationship between stability of the individual arene cluster molecules and that of the same molecules in the solid state has been addressed in terms of the relative crystal cohesion. The benzene ligands in apical coordination appear to be preferentially involved in C$-$H$\cdots$O$-$C interactions with respect to the facial ligands, as shown in Figure 6 for Ru$_6$C(CO)$_{11}$(η^6-C$_6$H$_6$)($\mu_3-\eta^2$:η^2:η^2-C$_6$H$_6$).[48]

Thus far we have focused our attention on the type of donor. We have provided several examples of (M)C$-$H systems acting as donors in the formation of hydrogen bonds of the C$-$H$\cdots$O type. We have also shown how the donor capacity depends on the hybridization state of the C-donor, on geometry and bonding modes of ligands, hence on the electronics of the coordination mode (whether μ_2, μ_3, η^1, η^5 or η^6) which, in turn, depends on the type of metal atom. We have shown that most common organic-type ligands, which abound in organometallic chemistry, provide a profusion of these weak donors. Little has been said on the type of acceptor groups. Most commonly the acceptor site is provided by the CO-ligand, which, thanks to the possibility of varying the polarity on the O-atom by changing the coordination mode, affords a very versatile basic site for C$-$H$\cdots$O bonding.

The basicity of CO and its participation in hydrogen bonding interaction has been recently reviewed and will not be discussed in the context of this article. The interested reader is referred to refs. 38 and 49.

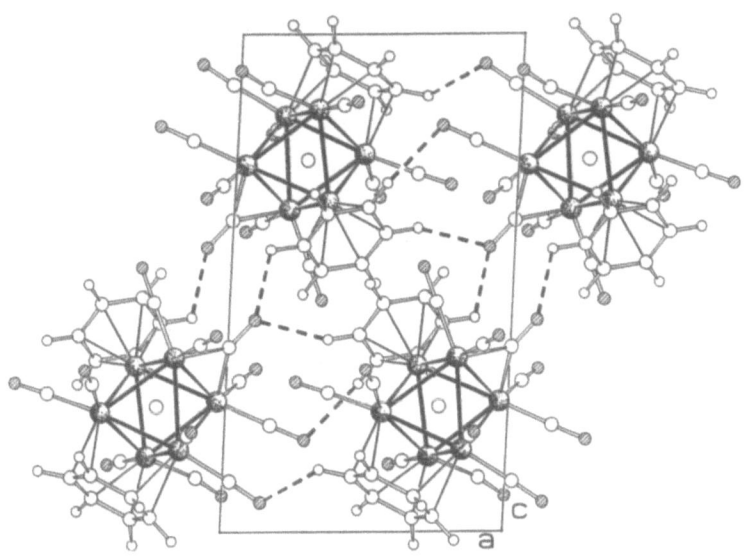

Figure 6. C−H···O hydrogen bonds in crystalline $Ru_6C(CO)_{11}(\eta^6-C_6H_6)(\mu_3-\eta^2:\eta^2:\eta^2-C_6H_6)$.

5. ORGANOMETALLIC M−C-H···O HYDROGEN BONDS TO WATER

Another common C−H···O hydrogen bond acceptor is water, which is often found as co-crystallization solvent in organometallic solids.[50] The water oxygen is a soft acceptor, comparable in strength to carbon monoxide or to the organic chetonic CO . The role of C−H···O interactions in satisfying the hydrogen bond requirements of water molecules has been demonstrated by studying hydrated crystals.[51] It has been shown that water oxygens can coordinate a large number of C−H donors while the number of OH donors from neighboring water molecules is confined to a maximum of four. A fundamental role in the hydrogen bonding of water molecules is played by the repulsions between the oxygen atoms.[52] This is probably the reason why four-fold and higher coordination is possible when some of the donors are CH groups while it is extremely rare with only water donors. The presence of a number of H···O(water) interactions higher than the number of available accepting sites (the O-atom lone-pairs) has been taken as indicative of the limited directional requirements of the C−H···O bonds and of their chiefly electrostatic nature.

C−H···O interactions involving water molecules have been recently shown to play a key role in the stabilization of crystalline organometallic hydrated hydroxides of the cations $[(C_6H_6)_2Cr]^+$ and $[(C_5H_5)_2Co]^+$. Dibenzene chromium hydroxide crystallizes with three water molecules per formula unit, i.e. as $[(C_6H_6)_2Cr]^+[OH]^-\cdot3H_2O^{14a,b}$ while cobalticinium hydroxide carries four water molecules.[53]

In crystalline $[(C_6H_6)_2Cr]^+[OH]^-\cdot3H_2O$, water and −OH groups form a slightly puckered hexagonal network with the O-atoms hydrogen bonded to three neighbor oxygens. The interlayer link is provided by three C−H···O interactions between each crystallographically independent oxygen atom and the benzene ligands above and below the hydrated layers. Each water molecule, besides donating two hydrogen bonds to neighboring molecules, accepts four donations, one of the O−H···O type and three C−H···O bonds. The hydrogen bonds distribution around the two crystallographically independent O-atoms is shown in Figure 7. Puckered, near hexagonal water networks of the type observed in crystalline $[(C_6H_6)_2Cr]^+[OH]^-\cdot3H_2O$ are present in ice[54] and in some inorganic hydroxides.[55]

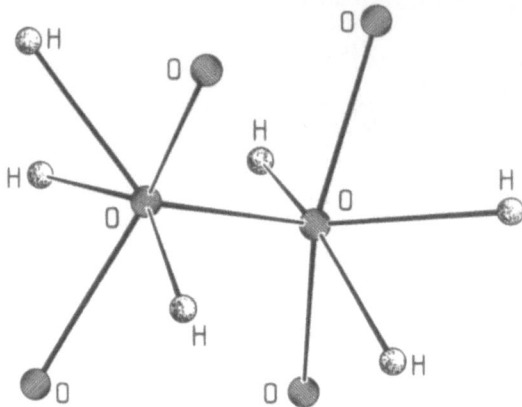

Figure 7. C−H···O and O−H···O hydrogen bonding around the two crystallographically independent water oxygen atoms in crystalline $[(C_6H_6)_2Cr]^+[OH]^-\cdot3H_2O$.

6. CHARGE ASSISTED C−H···O INTERACTIONS IN SALTS OF TRANSITION METAL CLUSTER ANIONS

Charge assistance has been discussed as a means to reinforce otherwise weak interactions. Charge assisted C−H···O bonds, as well as negatively charged O−H···O bonds, have been utilized to form supramolecular aggregates between organometallic cations and polycarboxylic organic acids.[14] Each time the donor group is carried by a cation and the acceptor group is carried by an anion, the weak electrostatic nature of the C−H···O hydrogen bond is strengthened by the different polarity of the two ions. In some cases this results in the formation of true ion pairs.

In the course of the investigation of the family of tetrahedral organometallic clusters derived from the binary carbonyls $[M_4(CO)_{12}]$ (M = Co, Rh, Ir) we have come across several cases of anionic clusters.[56] Cluster anions are invariably crystallized with large organic-type counterions, such NMe_4^+, PPh_4^+, PPN^+, etc. All these typical cations carry a large number of potential C−H donors. We have ascertained that C−H···O intermolecular bonds are appreciably shorter in crystalline salts than in the crystals formed by neutral species. In crystalline salts the interactions are between the CO-ligands of the anions and the H-atoms belonging to the cations and appear to be strengthened by the additional polarization arising from the net ionic charge delocalized over the clusters or the organic counterions.

Crystalline $[HIr_4(\mu_2-CO)_2(CO)_9][P(CH_2Ph)Ph_3]$,[56] for example, resents some very short C−H···O intermolecular bond involving the bridging CO. The shortest interaction [2.191 Å] is with the benzylic hydrogen, whereas the shortest distance from a phenyl hydrogen is 2.281 Å. The bridging CO is actually involved in a trifurcated interaction with the cation, as shown in Figure 8. Similarly, the bromide cluster $[Ir_4(\mu-CO)_3(CO)_8Br][PPh_4]$[56] shows the presence of a very short C−H···O intermolecular bond of 2.273 Å between one phenyl H-atom and a bridging CO. In crystalline $[Ir_4(\mu-CO)_3(CO)_8(COOMe)][NMe_2(CH_2Ph)_2]$[56] the shortest interaction [2.378 Å] involves a terminal ligand, while the bridging CO's participate in polyfurcated hydrogen bonds.

Another example, involving cobalt rather than iridium, is provided by crystalline $[Co_4(\mu-CO)_3(CO)_8COMe][PPh_4]$.[56] In this crystal, however, the shortest C−H···O interaction (2.385 Å) is established by the carboxymethyl group and the PPh_4^+ cation hydrogen atoms.

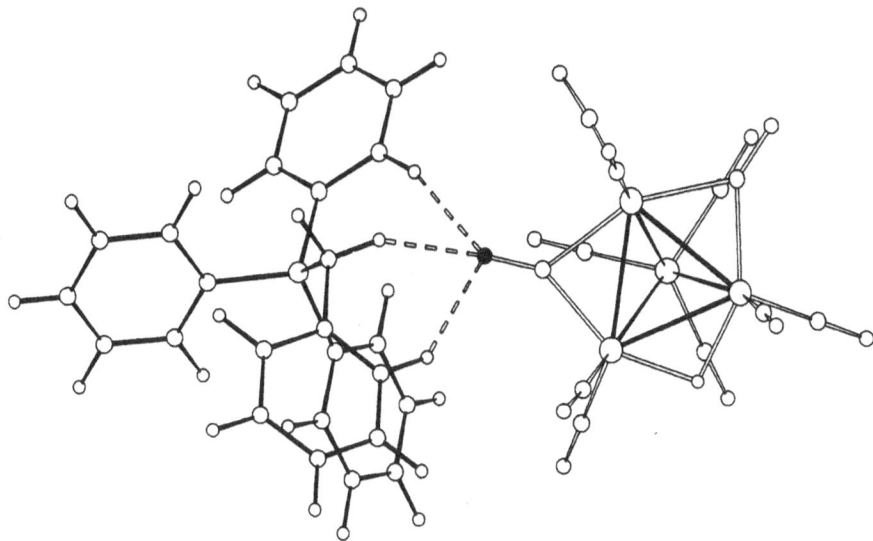

Figure 8. Charge assisted C−H···O bonds between anion and cation in crystalline [HIr$_4$(μ$_2$−CO)$_2$(CO)$_9$][P(CH$_2$Ph)Ph$_3$].

In summary, although crystalline salts formed of cluster anions and organic cations behave essentially as binary or ternary co-crystals in which particles are packed according to close packing rules, the ionic charge has the effect of "piloting" amongst the large number of possible alternative ways to construct these co-crystals towards those which also optimize the intermolecular hydrogen bondings.[57] The preference for hydrogen bonds with the bridging ligands with respect to terminal ones is not a hard and fast rule, rather a general tendency of complexes of the type mentioned above. When terminal ligands are preferred, it may well be because the maximization of the number of hydrogen bonds is more important than the type of carbonyl involved.

7. CONCLUSIONS

This review article has been devoted to a survey of the occurrence of C−H···O hydrogen bonding in organometallic crystals. The nature of this weak interaction between a polarized C−H donor and an oxygen acceptor has not been discussed in details. Rather we have taken advantage of the vast knowledge accumulated in the neighboring field of organic chemistry to focus on those features that are distinctive of organometallic species.

We have shown that ligand coordination to metal centers affords a great variety of different (new) C-H donor systems, some of which do not have an organic counterpart. Our observations can be summarized as follows:

(i) The donor capacity of C−H systems bound to metal via σ-bonds follows roughly the order μ$_3$−CH > μ$_2$−CH$_2$ > η^1−CH$_3$ which corresponds to the organic acidity scale C(sp)−H > C(sp^2)−H > C(sp^3)−H.

(ii) For C−H systems bound to metal centers in π-mode, as in the cases of the ligands η^5−C$_5$H$_5$, η^6−C$_6$H$_6$ and other carbocycles bound in delocalized fashion, or in the cases of alkenes and alkynes participate in hydrogen bonding interactions, the metric of these bonds corresponds roughly to that of C(sp^2)−H systems. The electronic nature and the oxidation state of the metal atom(s) may affect the acidity of the C−H bond, but this aspect has not yet been systematically studied.

(iii) In terms of acceptor capacity, the commonly found CO-ligand behaves very much like a tunable base as the basicity of the O-atom follows the order $\mu_3-CO > \mu_2-CO > \eta^1-CO$. The CO-ligand is present in large number on high nuclearity complexes such as carbonyl clusters. The combination of a large number of potential donors (e.g. metal co-ordinated arene ligands) and of a large number of CO-acceptors allows formation of extensive networks of C−H$\cdots$O interactions. These bonds lead to highly cohesive structures which can then be exploited to stabilize otherwise unstable complexes in the solid state.

(iv) The water molecule, often found as solvent of cocrystallization or as ligand in coordination complexes, can accept hydrogen bonds from C−H systems. These interactions are comparable in length (hence presumably in strength) to those established with CO-acceptors. Furthermore the water oxygen can afford hydrogen bonding with larger number of C−H donors than of O−H donors thus favoring interactions with organometallic molecules or ions that carry only C−H donors.

(v) Charge assistance, i.e. the different ionic charge carried by the potential C−H donor (as in the cations $[(C_6H_6)_2Cr]^+$ and $[(C_5H_5)_2Co]^+$ discussed above) and by the acceptor oxygen atoms can be advantageous to strengthening the C−H$\cdots$O bonding between cation donors and anion acceptors. This charge-assistance effect is also observed with carbonyl anions, where the basicity of the O-atom of the CO ligands is reinforced by the negative charge thus leading to the formation of relatively strong H-bonds with C−H donors belonging to organic counterions used for crystallization.

In summary, C−H$\cdots$O bonding is pervasive in organometallic crystals. Conventional strong donors, such as the −COOH, −OH, −CONRH etc. are not, on the other hand, often available in organometallic crystals for the simple reason that these groups, being strong Lewis bases, tend to interact directly with empty *d*-orbital on the metal atoms giving rise to dative bonds and complexation. Undoubtedly, in the presence of strong competitors the C−H$\cdots$O bond is sacrificed first, but when these strong competitors are absent the C−H$\cdots$O bond may display all its potentialities. It can be envisaged that in the near future researchers may realize that it is possible to *compensate weakness with number* in devising crystal synthesis strategies. One way is that of utilizing a large number of C−H donors of tunable polarity and/or a large number of O-acceptors of tunable basicity. This is possible with organometallic species for which it is often desirable (at times compulsory) to avoid the use of strongly acidic or basic groups. Crystal engineers of organometallic systems may exploit weak bonds to make materials containing metal atoms in predefined oxidation and spin states within predefined crystal structures.

Acknowledgment

Much of what we have learned about hydrogen bond comes from the fruitful collaboration with Professor Gautam Desiraju. The work discussed here has seen the contribution of numerous collaborators who share the authorship of the papers quoted throughout this article. We acknowledge financial support by the University of Bologna (project *"Intelligent Molecules and Molecular Aggregates"*, 1995), by MURST, and by bilateral exchange projects with Portugal (CNR-JINICT), and Germany (Vigoni Program, DAAR-CRUI), and the Erasmus student exchange program "Crystallography".

REFERENCES

1. (a) D. Braga and F. Grepioni, *J. Chem. Soc. Chem. Commun.* p. 571 (1996). (b) J. Hulliger, *Angew. Chem. Int. Ed. Engl.* 33:143 (1994). (c) C. L. Bowes and G. A. Ozin, *Adv. Mater.* 8:13 (1996). (d) G. A. Ozin, *Acc. Chem. Res.* 30:17 (1997). (e) M. J. Zaworotko, *Nature* 386:220 (1997). (f) O. M. Yaghi, L. Guangming and H. Li, *Nature* 378:703(1995). (g) P. Ball, *Nature* 381:648 (1996).

2. (a) A. Gavezzotti, *Acc. Chem. Res.* 27:309 (1994). (b) H. R. Karfunkel and R. J. Gdanitz, *J. Comput. Chem.* 13:1171 (1992). (c) R. J. Gdanitz, *Chem. Phys. Lett.* 190:391 (1992). (d) S. J. Maginn, *Acta Cryst.* A52:C79 (1996). (e) "Theoretical Aspects and Computational Modeling of the Molecular Solid State", A. Gavezzotti, ed., Wiley, Chichester (1997). (f) A. Gavezzotti, *in*: "Current Opinion in Solid State and Materials Science", A. K. Chitham, H. Inokuchi and J. M. Thomas, eds., vol. 1:501 (1996).

3. (a) O. Khan. "Molecular Magnetism", VCH Publishers, New York (1993). (b) O. Khan, *in*: "Inorganic Materials", D. W. Bruce and D. O'Hare, eds., Wiley, Chichester, U.K. (1992). (c) D. Gatteschi, *Adv. Mater.* 6:635 (1994).

4. (a) J. M. Williams, H. H. Wang, T. J. Emge, U. Geiser, M. A. Beno, P. C. W. Leung, K. Douglas Carson, R. J. Thorn, A. J. Schultz and M. Whangbo, *Progr. Inorg. Chem.* 35:218 (1987). (b) J. M. Williams, J. R. Ferraro, R. J. Thorn, K. D. Carlson, U. Geiser, H.-H. Wang, A. M. Kini and M.-H. Whangbo. "Organic Superconductors (Including Fullerenes): Synthesis, Structure, Properties and Theory", Prentice Hall, Englewood Cliffs, NJ (1992).

5. (a) J. S. Miller and A. J. Epstein, *Angew. Chem. Int. Ed. Engl.* 33:385 (1994). (b) J. S. Miller and A. J. Epstein, *Chem. Eng. News* 73:30 (1995).

6. (a) S. R. Marder, *Inorg. Mater.* p. 115 (1992). (b) N. J. Long, *Angew. Chem. Int. Ed. Engl.* 34:21 (1995). (c) T. J. Marks and M. A. Ratner, *Angew. Chem. Int. Ed. Engl.* 34:155 (1995). (d) D. R. Kanis, M. A. Ratner and T. J. Marks, *Chem. Rev.* 94:195 (1994).

7. (a) M. C. Etter, *Acc. Chem. Res.* 23:120 (1990). (b) G. R. Desiraju, *Angew. Chem. Int. Ed. Engl.* 34:2311 (1995).

8. (a) G. R. Desiraju. "Organic Solid State Chemistry", Elsevier, Amsterdam (1987). (b) G. R. Desiraju. "Crystal Engineering: The Design of Organic Solids", Elsevier, Amsterdam (1989).

9. (a) A. D. Burrows, C.-W. Chan, M. M. Chowdry, J. E. McGrady and D. M. P. Mingos, *Chem. Soc. Rev.* p. 329 (1995). (b) S. Subramanian and M. J. Zaworotko, *Coord. Chem. Rev.* 137:357(1994).

10. (a) G. A. Jeffrey and W. Saenger. "Hydrogen Bonding in Biological Structures", Springer-Verlag, Berlin (1991). (b) C. B. Aakeröy and K. R. Seddon, *Chem. Soc. Rev.* p. 397 (1993). (c) L. Brammer, D. Zhao, D. T. Ladipo and J. Braddock-Wilking, *Acta Cryst.* B51:632 (1995).

11. M. Meot-Ner (Mautner), *J. Am. Chem. Soc.* 106:1257 (1984).

12. M. Meot-Ner (Mautner) and L.W. Sieck, *J. Am. Chem. Soc.* 108:7525 (1986).

13. (a) G. Gilli, F. Bellucci, V. Ferretti and V. Bertolasi, *J. Am. Chem. Soc.* 111:1023 (1989). (b) V. Bertolasi, P. Gilli, V. Ferretti and G. Gilli, *J. Am. Chem. Soc.* 1113:4917 (1991). (c) G. Gilli, V. Bertolasi, V. Ferretti and P. Gilli, *Acta Crystallogr.* B49:564 (1993) (d) V. Bertolasi, P. Gilli, V. Ferretti and G. Gilli, *Chem. Eur. J.* 2:925 (1996).

14. (a) D. Braga, F. Grepioni, J. J. Byrne and A. Wolf, *J. Chem. Soc. Chem. Commun.* p. 1023 (1995). (b) D. Braga, A. L. Costa, F. Grepioni, L. Scaccianoce and E. Tagliavini, *Organometallics* 15:1084 (1996). (c) D. Braga, A. Angeloni, F. Grepioni and E. Tagliavini, *J. Chem. Soc. Chem. Commun.* p. 1447 (1997).

15. (a) G. R. Desiraju, *Acc. Chem. Res.* 19:222 (1986). (b) G. R. Desiraju, *Acc. Chem. Res.* 29:441 (1996). (c) T. Steiner, *Chem. Commun.* p. 727 (1997).

16. (a) A. Allerhand and P. von Ragué Schleyer *J. Am. Chem. Soc.* 85:1715 (1963). (b) T. Steiner, *Cryst. Rev.* 6:1 (1996).

17. D. Mootz and A. Deeg, *J. Am. Chem. Soc.* 114:5887 (1992).

18. M. A. Viswamitra, R. Radhakrishnan, J. Bandekar and G. R. Desiraju, *J. Am. Chem. Soc.* 115:4868 (1993).

19. Z. Berkovitch-Yellin and L. Leiserowitz, *J. Am. Chem. Soc.* 104:4052 (1982)

20. C. V. Sharma and G. R. Desiraju, *J. Chem. Soc. Perkin Trans.* 2:2345 (1994).

21. R. O. Gould, A. M. Gray, P. Taylor and M. D. Walkinshaw, *J. Am. Chem. Soc.* 107:5921 (1985)

22. D. Braga and F. Grepioni, *Coord. Chem. Rev.* in press

23. F. H. Allen, J. E. Davies, J. J. Galloy, O. Johnson, O. Kennard, C.F. Macrae and D. G. Watson, *J. Chem. Inf. Comp. Sci.* 31:204 (1991).

24. D. Braga and F. Grepioni, *Acc. Chem. Res.* 27:51 (1994).

25. (a) D. Braga, A. L. Costa, F. Grepioni, L. Scaccianoce and E. Tagliavini, *Organometallics* 15:1084 (1996). (b) D. Braga, A. L. Costa, F. Grepioni, L. Scaccianoce and E. Tagliavini, *Organometallics* 16:2070 (1997).

26. D. Braga, F. Grepioni and G. R. Desiraju, *J. Organomet. Chem.* in press.

27. D. Braga, F. Grepioni, E. Tedesco, K. Biradha and G. R. Desiraju. *Organometallics* 15:2692 (1986).

28. F. A. Cotton and G. Wilkinson, "Advanced Inorganic Chemistry", fifth edition, Wiley, New York (1993).

29.   (a) M. A. Beno, J. M. Williams, M. Tachikawa and E. L. Muetterties, *J. Am. Chem. Soc.* 102:4542 (1980).
      (b) M. A. Beno, J. M. Williams, M.Tachikawa and E. L. Muetterties, *J. Am. Chem. Soc.* 103:1485 (1981).
      (c) H. Wadepohl, D. Braga and F. Grepioni, *Organometallics*, 14:24 (1995).

30.   (a) D. A. Clemente, B. Rees, G. Bandoli, M. C. Biagini, B. Reiter and W. Herrmann, *Angew. Chem. Int. Ed. Engl.* 20:887 (1981). (b) W. A. Herrmann, *Adv. Organomet. Chem.* 20:159 (1982). (c) M. I. Altbach, F. R. Fronczek and L. G. Butler, *Acta Crystallogr.* C48:644 (1992). (d) D. C. Miller and T. B. Brill, *Inorg. Chem.* 17:240 (1978).

31.   (a) P. Hofmann, *Angew. Chem. Int. Ed. Engl.* 18:554 (1978). (b) D. C. Calabro, D. L. Lichtenberger and W. A. Herrmann, *J. Am. Chem. Soc.* 103:6852 (1981). (c) B. E. Bursten and R. H. Cayton, *J. Am. Chem. Soc.* 108:8241 (1986). (d) B. E. Bursten and R. H. Cayton, *J. Am. Chem. Soc.* 109:6053 (1987).

32.   D. Braga, F. Grepioni, E. Tedesco, S. Gebert and H. J. Wadepohl, *Chem. Soc. Dalton Trans.* 1997, 1727.

33.   P. Leung, P. Coppens, R. K. McMullan and T. F. Koetzle, *Acta Crystallogr.* B37:1347 (1981).

34.   (a) D. A. Clemente, M. C. Biagini, B. Rees and W. A. Herrmann, *Inorg. Chem.* 21:3741 (1982).

35.   (a) R. F. Bryan, P. T. Greene, M. J. Newlands and D. Field, *J. Chem. Soc. A* p. 3068 (1970); (b) R. F. Bryan and P. T. Greene, *J. Chem. Soc. A* p. 3064 (1970).

36.   (a) S. Aime, M. Botta, R. Gobetto amd A. Orlandi, *Magn. Res. Chem.* 28:S52 (1990). (b) D. Braga, C. Gradella and F. Grepioni, *J. Chem. Soc. Dalton Trans.* p. 1721 (1989).

37.   T. Steiner, *J. Chem. Soc. Chem. Commun.* p. 101 (1994).

38.   D. Braga, F. Grepioni, K. Biradha, V. R. Pedireddi and G. R. Desiraju, *J. Am. Chem. Soc.* 117:3156 (1995).

39.   R. D. Adams, D. M. Collins and F. A. Cotton, *J. Am. Chem. Soc.* 96:749 (1974).

40.   M. D. Curtis and W. M. Butler, *J. Organomet. Chem.* 155:131 (1978).

41.   D. Braga, P. J. Dyson, F. Grepioni and B. F. G. Johnson, *Chem. Rev.* 94:1585 (1994).

42.   D. Braga, F. Grepioni, B. F. G. Johnson, J. Lewis, C. E. Housecroft and M. Martinelli, *Organometallics* 10:1260 (1991).

43.   (a) L. Hoferkamp, G. Rheinwald, H. Stoeckli-Evans and G. Süss-Fink, *Helv. Chim. Acta* 75:2227 (1992). (b) S. Rossi, K. Kallinen, J. Pursiainen, T. T. Pakkanen and T. A. Pakkanen, *J. Organomet. Chem.* 440:367 (1992).

44.   D. Braga, F. Grepioni, H. Wadepohl, S. Gebert, M. J. Calhorda amd L. F. Veiros, *Organometallics* 14:5350 (1995).

45.   (a) D. Braga and F. Grepioni, *J. Chem. Soc. Dalton Trans.* p. 1223 (1993). (b) C. Renouard, G. Rheinwald, H. Stoeckli-Evans; G. Süss-Fink, D. Braga and F. Grepioni, *J. Chem. Soc. Dalton Trans.* p. 1875 (1996).

46.   D. Braga, F. Grepioni, P. Sabatino, P. J. Dyson, B. F. G. Johnson, J. Lewis, P. J. Bailey, P. R. Raithby and D. J. Stalke, *Chem. Soc. Dalton Trans.* p. 985 (1993).

47.   (a) M. F. Gomez-Sal, B. F. G. Johnson, J. Lewis, P. R. Raithby and A. H. Wright, *J. Chem. Soc. Chem. Commun.* p. 1682 (1985); (b) P. J. Dyson, B. F. G. Johnson, J. Lewis, M. Martinelli, D. Braga and F. Grepioni, *J. Amer. Chem. Soc.* 115:9062 (1993). (c) R. D. Adams and W. Wu, *Polyhedron* 2:2123 (1992).

48.   D. Braga, P. J. Dyson, F. Grepioni, B. F. G. Johnson and M. Calhorda, *J. Inorg. Chem.* 33:3218 (1994).

49.   D. Braga and F. Grepioni, *Acc. Chem. Res.* 30:81 (1997).

50.   D. Braga and F. Grepioni, *Comments Inorg. Chem.* 19:185 (1997).

51.   H. F. Savage and J. L. Finney, *Nature* 322:717 (1986).

52.   T. Steiner and W. Saenger, *J. Am. Chem. Soc.* 115:4540 (1993).

53.   D. Braga, A. Angeloni and F. Grepioni, unpublished results.

54.   M. Falk and O. Knop. "Water. A Comprehensive Treatise", F. Franks, ed., Plenum Press, New York (1973), vol. 2, pp. 55.

55.   A. F. Wells. "Structural Inorganic Chemistry", Clarendon Press, Oxford (1984).

56.   D. Braga, J. J. Byrne, M. J. Calhorda and F. Grepioni, *J. Chem. Soc. Dalton Trans.* p. 3287 (1995) and references therein.

57.   (a) D. Braga, F. Grepioni, P. Milne and E. Parisini, *J. Am. Chem. Soc.* 115:5115 (1993). (b) D. Braga, F. Grepioni and E. Parisini, *J. Chem. Soc. Dalton Trans.* p. 287 (1995).

THE IMPORTANCE OF INTRA- AND INTERMOLECULAR WEAK BONDS IN TRANSITION METAL COORDINATION COMPOUNDS

Peter Comba

Anorganisch-Chemisches Institut der Universität Heidelberg
Im Neuenheimer Feld 270
D–69120 Heidelberg
Germany

1. INTRODUCTION

There is an increasing interest in supramolecular systems involving inorganic and organic compounds with specific properties that may lead, eventually, to novel materials and devices and to challenging advances in materials science and biochemistry. Supramolecular chemistry is the "chemistry beyond the molecule", i.e. the "chemistry controlled by non-covalent intermolecular forces".[1] Of specific interest are self-assembled arrays of molecules and self-organized supramolecular systems. The general textbook definition describes a molecule as a "well defined assembly of atoms bound to each other", as the "smallest unit of a pure compound with the specific chemical properties of the corresponding bulk material".[2] These definitions of molecules and supramolecular systems are as intuitive as they are ambiguous. Is the array of water molecules in beautiful ice flowers a supramolecular assembly, is the crystallization of sodium chloride a self-organization process? Challenging and truly interesting supramolecular systems have properties that are different from those of the constituent "molecules". Where then is the limit between a molecule and a supramolecular system, which of the interactions are intra- and which are intermolecular? When does an interaction cease to be a bond and start to be a nonbonded interaction? In order to circumvent problems of this kind, I will often use the term "system" instead of "molecule", and I will not distinguish between inter- and intramolecular bonds and interactions. The aim of the present chapter is to demonstrate the importance of interactions in coordination compounds that are not always appreciated as bonds. The examples cover structural and thermodynamic aspects, as well as spectroscopy and reactivity. Since some of these studies are discussed with the help of molecular modeling, a few general and important aspects of modeling related to intermolecular interactions are presented first.

2. MOLECULAR MODELING

The environment is often neglected in molecular modeling studies, i.e. interactions to counter ions and solvent molecules in solution, as well as the crystal lattice in the solid state are often not explicitly included when structures, thermodynamic and spectroscopic properties of molecular compounds are computed. In spite of this, there are many reasons why acceptable or excellent agreement between experimentally observed and computed properties may be, and often are, obtained. For force field based methods, the primary reason is that molecular mechanics is an interpolative method, and the parameterization is generally based on solution or solid state data. That is, the environment is implicitly included in the model and, therefore, the computed structures, spectra and strain energies of "interactionless molecules" are those of the species in an averaged, isotropic crystal lattice (or solution). That is, these calculations are not related to gas phase molecules.[3] Therefore, computed structures and molecular properties of high accuracy may be expected if specific effects that are not included in the model (ion-pairing, solvation, entropic effects) are constant or highly correlated to the strain energy within the series of compounds considered.[3]

There are force field based methods that allow the including of crystal lattices and the environment in solution (see various other chapters in this book, references 3, 4 and publications cited therein). The main reason for not generally applying these techniques is their computational expense. It is important to note here that, if a program is used which minimizes the molecule in its environment, a force field different to that used for the computation of interactionless molecules has to be chosen. That is, a force field based on gas phase data (structures – e.g. ab-initio MO calculations, thermodynamics, spectroscopy) will produce gas phase structures when interactionless molecules are computed, and it will produce solid state or solution structures if an array of molecules in a crystal lattice or the molecule in its solvent sheath are minimized. A conventional force field, based on solid state data (e.g. X-ray crystal structures) already has these environmental influences included and should not be used for minimizing a crystal lattice or solvated molecules. The reason is that the interactions that are neglected when minimizing isolated molecules are nonbonded interactions (van der Waals and electrostatic interactions), and these are attractive at long distances. Thus, an isolated molecule in a lattice must experience some pressure that may lead to a contraction of the structure. This force is built-in in force fields for interactionless molecules, based on solid state data.

Computed structures of isolated molecules are often more symmetrical than experimental data based on crystal structures. (Note that this is not necessarily so. For example, $[M(NH_3)_6]^{n+}$ does not refine to a structure that is as symmetrical as one naively might expect. This is due to the coincidence of the three-fold axes of the ammonia donors with the four-fold axes of the MN_6 core.) Distortions in experimental structures are therefore often related to "crystal lattice effects". This may not always be the only and correct interpretation, and a thorough analysis of these effects must involve minimization of the whole lattice. Note also that the fact that a solid state and a solution structure (e.g. by NMR/NOE analysis) are similar (same conformation) and in agreement with a computed structure does not prove the absence of crystal lattice effects since the type of interactions in solution and in the solid may be similar, i.e. mainly electrostatic (ion-pairing, H-bonding) for water soluble and predominantly van der Waals for hydrophobic compounds.

3. LIGAND FIELD SPECTROSCOPY: FOUR-, FIVE- AND SIX- COORDINATE COPPER(II) COMPOUNDS

Copper(II) compounds, in particular copper(II) tetraamines, usually have four tightly bound donors in a square planar arrangement with one or two axial ligands at longer distances (4 + 1 or 4 + 2 coordination geometries). There is an inverse correlation involving the in-plane and the axial bond distances, i.e. strong in-plane bonding generally leads to weak axial interactions.[5] Also, the axial bonds in five-coordinate systems are, as expected, generally shorter than in structures with two axial donors (~ 2.3 Å vs. ~ 2.6 Å for axial OH_2).

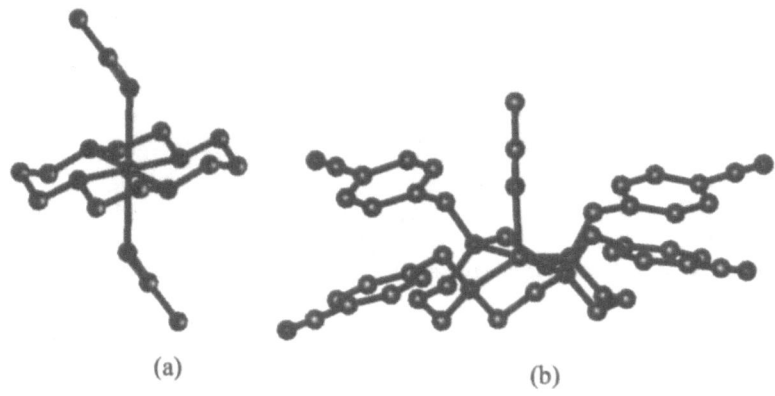

(a)   (b)

Figure 1. Copper(II) compounds of 14-membered tetraazamacrocyclic ligands and axially coordinated acetonitrile. (a) [Cu(cyclam)(NCCH$_3$)$_2$]$^{2+}$ (cyclam = 1,4,8,11-tetraazacyclotetradecane);[6] (b) [Cu(L^1)(NCCH$_3$)]$^{2+}$ (L^1 = 1,4,8,11-tetra-4-cyanobenzyl-1,4,8,11-tetraazacyclotetradecane).

Note, however, that there are some examples of copper(II) tetraamines with exceptionally short axial bonds. An interesting recent example is Cu(L^1)(NCCH$_3$)$_n^{2+}$, where L^1 is a cyclam derivative (cyclam = 1,4,8,11-tetraazacyclotetradecane), and n is 1, 2 (see Fig. 1). With L^1 = cyclam and $n = 2$ the configuration of the coordinated macrocycle is R*,R*,S*,S* (configuration of the coordinated amines, trans III, Fig. 1a),[6] with N-substituted derivatives the preferred configuration is R*,S*,R*,S* (trans I).[7] With L^1 = 1,4,8,11-tetra-4-cyanobenzyl-1,4,8,11-tetraazacyclotetradecane copper(II) forms a deep green compound with one coordinated CH$_3$CN donor (Fig. 1b).[8] The in-plane Cu–N bonds of this species are slightly asymmetrical and significantly longer than in the parent compound with L^1 = cyclam (2.15, 2.10, 2.17, 2.08 Å vs. 2.02 Å; the average Cu–N distance in other copper(II) compounds with 14-membered tetraazamacrocyclic ligands is 2.01 Å). The axial interaction to CH$_3$CN in the green compound is extremely short compared to that in the parent compound (2.19 Å vs. 2.57 Å; the axial bonds to OClO$_3^-$ in six-coordinate copper(II) tetraamines are $\sim$ 2.6 Å; in other five-coordinate copper(II) complexes with substituted cyclam derivatives, leading to the trans I configuration, the Cu–O^{ax} distances are considerably longer than that observed here for Cu–NCCH$_3$, i.e. $\sim$ 2.4 Å vs. 2.19 Å).[9,10]

Ligand field spectroscopy is an excellent method to probe the weak interactions to axial donors in copper(II) coordination compounds,[11] and angular overlap model (AOM)[11,12] calculations are an elegant tool for the analysis and interpretation of the corresponding electronic and EPR spectroscopic data. The qualitative diagram of Fig. 2 visualizes the salient features (for copper(II) tetraamines metal-donor π interactions are negligible):

1. Strong in-plane interactions lead to high-energy ligand field transitions.

2. Strong σ-interactions to axial donors lead to a decrease in the energy of the ligand field transitions due to the weakening of the in-plane bonds (inverse correlation of in-plane and axial bonds[5]).

3. Angular distortions within the CuN$_4$ plane decrease the overlap between the donor σ orbitals and the metal d orbitals and, therefore, lead to a decrease in energy of the ligand field transitions.

An instructive example to demonstrate the influence of axial donor groups is that of the color of the two isomeric forms of Cu(L^2)$_2$·X$_2$, where L^2 is the chiral diamine ahaz (ahaz =

$$
\left\{
\begin{aligned}
E(d_{x^2-y^2}) &= 3e_\sigma \\
E(d_{z^2}) &= e_\sigma & +(2)e_\sigma^{ax} & -4e_{ds} \\
E(d_{xy}) &= (4e_\pi) \\
E(d_{xz}, d_{yz}) &= (2e_\pi) & +(2)e_\pi^{ax}
\end{aligned}
\right.
$$

$$\text{CuN}_4 \qquad \text{axial ligand(s)} \qquad ds\text{-mixing}$$

Figure 2. AOM d-orbital splitting of four- (4 + 1, 4 + 2) coordinate copper(II) tetraamines.

3-aminohexahydroazepine) and X is an anion (e.g. $OClO_3^-$): a purple ($\lambda_{max} = 518$ nm) five-coordinate complex is obtained with optically pure L^2 while an orange ($\lambda_{max} = 450$ nm) copper(II) tetraamine compound (four-coordinate) precipitates from a solution containing racemic L^2 (see Fig. 3).[13]

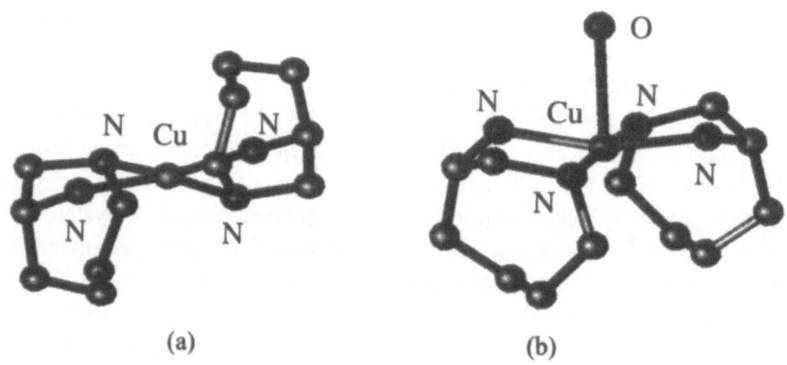

(a)   (b)

Figure 3. The two isomers of $[\text{Cu(ahaz)}_2\text{X}_2]$: (a) $[\text{Cu(R*-ahaz)(S*-ahaz)](ClO}_4)_2$[13] (orange), (b) $[\text{Cu(S*-ahaz)}_2\text{ (OClO}_3)]\text{ClO}_4$ (purple).[10]

AOM calculations of the two isomeric forms of $\text{Cu(L}^2)_2\cdot\text{X}_2$ and some related copper(II) tetraamines, using a constant set of electronic parameters (adjusted for variations in the M−L distances, i.e. $e_\sigma = f(1/r^6)$, where $r =$ M−L), demonstrate the dependence of the ligand field spectroscopic properties on the coordination geometry, in particular the correlation between the dd absorption and the bonding of axial donor groups (Table 1).[10]

4. REACTIVITY: THE FORMALDEHYDE/NITROETHANE TEMPLATE CONDENSATION

The copper(II) directed synthesis of the tetradentate ligands L^3, based on the Mannich condensation of amino acids with formaldehyde and nitroethane, may lead to three isomers, when starting with a racemic amino acid (Fig. 4). However, ^{1}H-NMR spectra of the metal-free ligands indicate that the reaction is fully stereoselective and leads to a product (structures (b) or (c)) that is different from that obtained from optically pure amino acids (structure (a)).[14]

Table 1. Experimentally determined and calculated (italics) structural and spectroscopic parameters (MM-AOM in brackets) of a series of copper(II) tetraamines [10]

Compound[a]	Cu–N^{av}(Å)		Cu–X^{av}(Å)	ϑ^{av}(deg)[b]	E_{xy} (cm^{-1})
[Cu(en)$_2$]$^{2+}$	2.028	2	2.593	0.0	19700 (*21100*)
	2.005		*2.522*	*0.0*	*20430*
[Cu(R*-ahaz)(S*-ahaz)]$^{2+}$	2.008	0		0.0	22210 (*21630*)
	2.016			*8.2*	*20160*
[Cu(S*-ahaz)$_2$]$^{2+}$	2.013	1	2.437	12.3	19300 (*19720*)
	2.014		*2.317*	*14.6*	*19900*
[Cu([13]N$_4$)]$^{2+}$	2.014	1	(2.507)[c]	28.0	18020 (*19530*)
	1.999		*2.195*	*21.6*	*18640*
[Cu([14]N$_4$)]$^{2+}$	2.006	2	2.558	0.3	20830 (*21200*)
	2.019		*2.549*	*1.1*	*20400*
[Cu([15]N$_4$)]$^{2+}$	2.037	2	2.512	11.5	18180 (*18840*)
	2.038		*2.493*	*17.2*	*18340*
[Cu([16]N$_4$)]$^{2+}$	2.029	2	2.643	35.3	17240 (*18080*)
	2.008		*2.682*	*42.5*	*17850*

[a]en = ethane-1,2-diamine; ahaz = 3-aminohexahydroazepine; [n]N$_4$: n-membered tetraazamacrocycle
[b]tetrahedral twist angle (square planar: $\vartheta = 0°$, tetrahedral: $\vartheta = 90°$)
[c]experiment: X = Cl; modeling: X = OH$_2$

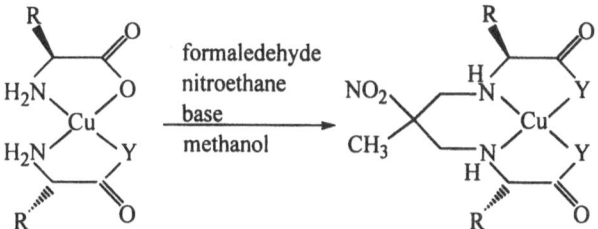

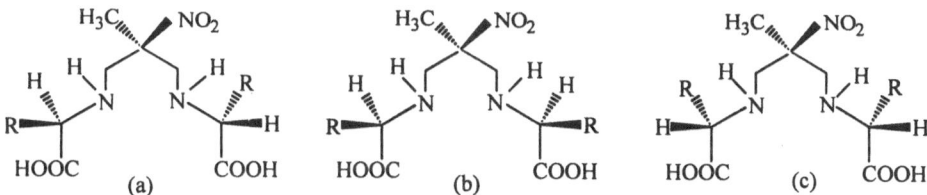

Figure 4. The stereoselective copper(II) directed synthesis of tetradentate ligands.

The proposed mechanism involves coordinated imines that react with the deprotonated α-methylene group of nitroethane to form the six-membered chelate ring. The face-selectivity that leads to structure (c) must involve the nitroethane molecule approaching the coordinated imines from the chromophore rather than from the periphery of the molecule (Fig. 5). This hypothesis is supported by three observations:[15]

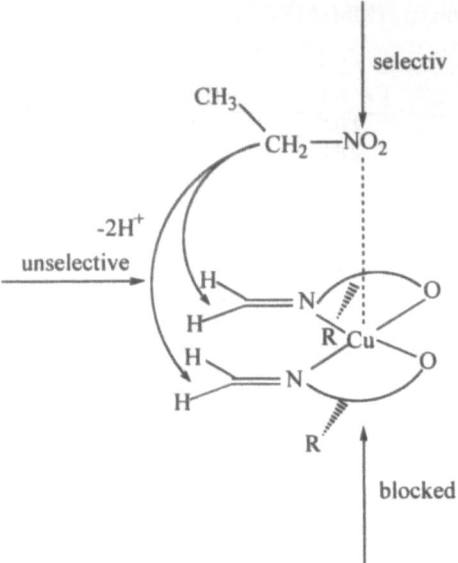

selectiv

CH₃

CH₂—NO₂

-2H⁺

unselective

blocked

Figure 5. Putative reason for the observed stereoselectivity.

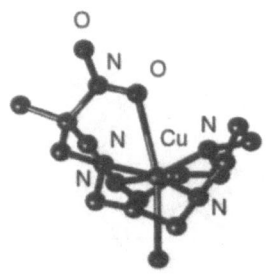

Figure 6. Experimentally determined structure of $[Cu([16]N_4)(OClO_3)]^{+}$[17], see also Table 1.

1. An X-ray crystallographic analysis identified the coordinated tetradentate ligand as that of structure (c) in Fig. 4.

2. Qualitative molecular mechanics calculations indicate an energy difference of at least $5\,\text{kJ}\,\text{mol}^{-1}$ between the putative intermediate, leading to the observed structure, and those, leading to structures (a) and (b) in Fig. 4 (note that the intermediate responsible for the selectivity might be different from the species shown in Fig. 5, and various structures, e.g. half-cyclized and carbinolamine-type intermediates lead to similar strain energy differences).

3. An analogous template reaction, based on $[Cu(en)_2]^{2+}$, that leads to the cyclam derivative dinemac (dinemac = 6,13-dimethyl-6,13-dinitro-1,4,8,11-tetraazacyclotetradecane) is not fully stereoselective when using an excess of aqueous formaldehyde, and it leads to an 80 % / 20 % mixture of trans- and cis-dinemac.[16] However, with paraformaldehyde, full stereoselectivity is observed, indicating that, in the initial reaction, water may compete with the axial coordination of the organic nitro group.[15]

Weak axial interactions to organic nitro substituents have been observed in related copper(II) compounds of macrocyclic ligands with pendent nitro groups[10,17] (Fig. 6; see last entry in Table 1). Also, there is infrared spectroscopic evidence for $[M(CH_3NO_2)_6]^{2+}$ (M = Mg[II], Mn[II], Fe[II], Co[II], Ni[II]).[18]

5. CONFORMATIONAL EQUILIBRIA: DINUCLEAR COPPER(II) AND NICKEL(II) COMPOUNDS

The bismacrocyclic ligand L^4 (see Fig. 7) has two identical cavities with two amine and two amide donors each. The latter are deprotonated upon coordination to copper(II) or nickel(II) cations. The experimentally determined, averaged Cu–N_{amine} distances are 2.032 Å, Cu–N_{amide} is 1.930 Å, Cu–OH_2 is 2.301 Å, Ni–N_{amine} is 1.900 Å, Ni–N_{amide} is 1.842 Å. As expected, the 13-membered macrocyclic rings are slightly too small for Ni[II] and more so for Cu[II]. Thus, the metal centers are out of the N_4 planes, and the angles between the N_{amide}–M–N_{amide} and the N_{amine}–M–N_{amine} planes indicate a small but significant folding of the macrocycle in the case of M = Ni[II] ($\vartheta = 13°$) and a substantial folding for M = Cu[II] ($\vartheta = 40°$).[19] Interestingly, there is a fundamental difference between the experimentally determined dicopper(II) and the dinickel(II) solid state structures: The former is stretched ($\phi_{Cu-C-C-Cu} = 180°$) while the latter is folded ($\phi_{Ni-C-C-Ni} = 55°$; see Fig. 7).[19]

The solution structure of the dicopper(II) compound, determined by the combination of EPR spectroscopy with force field calculations and the simulation of the EPR spectra (MM-EPR),[19,20] is identical to that in the solid state (Fig. 8).[19] The two interesting questions (i) what is the solution structure of the dinickel(II) compound and (ii) what is the reason for the striking structural difference between the dinickel(II) and the dicopper(II) compounds may be answered by molecular modeling. A preliminary account of this study[19] is given here.

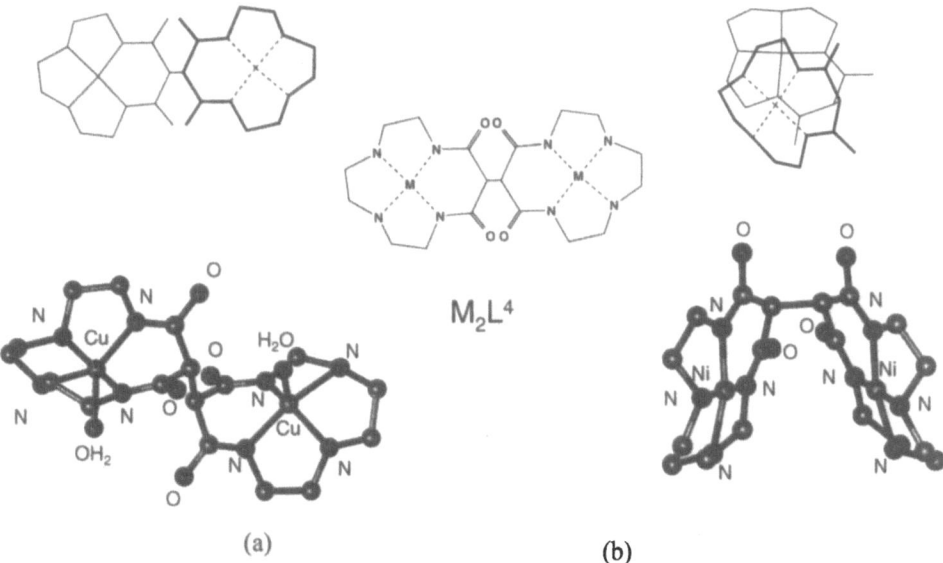

Figure 7. The two strikingly different structures of M_2L^4 for (a) M = Cu[II] and (b) M = Ni[II]; the pseudo-torsion angles ϕ are defined as M–$C_{bridgehead}$–$C'_{bridgehead}$–M'.

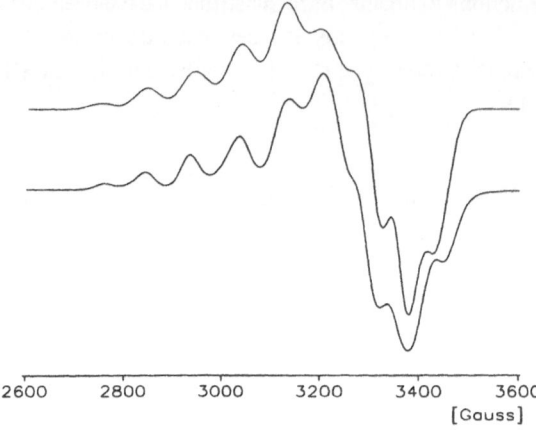

2600   2800   3000   3200   3400   3600

[Gauss]

Figure 8. Experimentally determined (bottom) and simulated (top) EPR spectra of Cu_2L^4.

The important feature emerging from the solid state structures (Fig. 7) is that a folded structure for the dicopper(II) compound is prevented by one axially coordinated water molecule to each of the two copper(II) centers. The important questions are (i) what is the reason for the relative stability of the folded structure for $M = Ni^{II}$ and (ii) why are the axial donors of the dicopper(II) compound not directed toward the periphery of the dinuclear system?

A possible driving force for the stabilization of the two observed structures is a difference in intermolecular forces (crystal lattice in the solid, solvation in solution, i.e. H-bonding and van der Waals interactions). Interestingly, the two dinuclear compounds crystallize with a number of crystal waters present (Cu^{II}: 14 H$\cdots$O interactions, $\sim$ 2.7 Å; Ni^{II}: 10 H$\cdots$O interactions, $\sim$ 2.2–3.0 Å). The extensive hydrogen bonding networks have been analyzed in both cases, and it emerges that, in total, the strength of the hydrogen bonding is similar for the two structures. Thus, hydrogen bonding in the solid and solvation energy differences in solution may not be a primary reason for the stabilization of the two structural motifs.

A conformational analysis of the dinickel(II) compound defines two distinct low energy structures with $\phi_{Ni-C-C-Ni} = 55°$ and $\phi_{Ni-C-C-Ni} = 180°$. The former is more stable by 20 kJ mol^{-1} and is similar to that observed in the solid state, the latter is similar to the conformation observed in the dicopper(II) compound. Our preliminary analysis does not involve electrostatic terms but an analysis of structurally related, charged bis-tetraamine dicopper(II) compounds (4+ cations; the dinuclear compounds studied here have two deprotonated amides per metal site and are therefore neutral) has indicated that electrostatic repulsion was of minor importance.[20c] The analysis of all strain energy terms of the two conformers points to van der Waals interactions as the force responsible for the stabilization of the folded structure. As noted in the introduction, these are stabilizing at long distances with an energy minimum at the distance of the sum of the van der Waals radii.

The strain energy versus $\phi_{Ni-C-C-Ni}$ plot (Fig. 9) shows two minima ($\phi_{Ni-C-C-Ni} = 55°$, $U_{strain} = -19$ kJ mol^{-1}; $\phi_{Ni-C-C-Ni} = 180°$, $U_{strain} = 0.0$ kJ mol^{-1}) and a maximum ("eclipsed" intermediate structure) at $\phi_{Ni-C-C-Ni} = 120°$ with a corresponding energy of $U_{strain} = 5.0$ kJ mol^{-1}.

Strain energy versus torsion angle curves for the dicopper(II) compound (four- and five-coordinate) are also included in Fig. 9. As expected, for five-coordinate copper(II), the only stable structure is the stretched form with a $\phi_{Cu-C-C-Cu}$ torsion of 180°. However, even for a hypothetical bis-four-coordinate dicopper(II) compound, the energy difference between the stretched and folded forms are much smaller than for nickel(II). Hence, the folded structure is only marginally more stable than the stretched form. (Note that this analysis is only based on steric effects, i.e. it does not include the loss of bonding energy due to dissociation

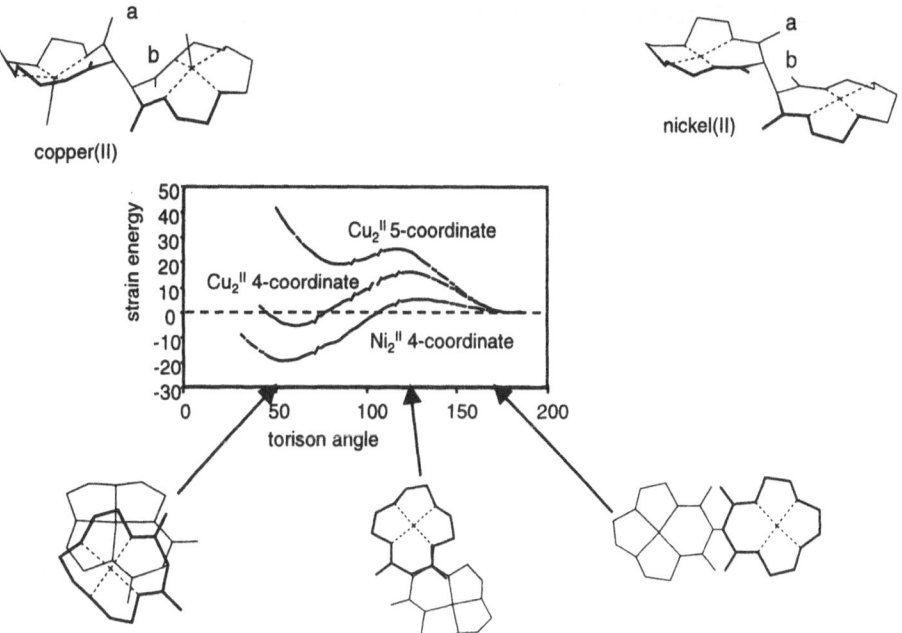

Figure 9. U_{strain} versus $\phi_{M-C-C-M}$ curves for M_2L^4 for $M = Ni^{II}$ and $M = Cu^{II}$ (4- and 5-coordinate).[19]

of two copper-water bonds.) This is due to the significant distortion of the CuN_4 fragment from planarity. Note that the $Cu\cdots Cu$ repulsion is underestimated in our model calculations which do not involve electrostatic effects (see above), i.e. the repulsion due to residual positive charges on the copper centers is not included here.

The question of why the out-of-plane bending of the chromophores is directed toward the neighboring copper(II) center ("convex" versus "concave" structure of the dinuclear complex) has not been addressed directly in our molecular mechanics calculations so far, and a thorough and more complete analysis, involving all possible configurations and conformations, is in progress.[19] However, the salient qualitative features are visualized in Fig. 9: Due to the planarity of the coordinated amide donor there is a significant bending of the amide oxygens out of the N_4 plane, exo to that of the copper(II) center. Thus, in the more planar structure of the dinickel(II) complex the $O^a_{amide}\cdots O^b_{amide}$ distances (see Fig. 9) are ~ 2.3 Å–2.8 Å, in the experimentally observed "convex" structure of the dicopper(II) compound the corresponding distances are ~ 3.8 Å–3.9 Å, while they are reduced to ~ 2.0 Å–2.3 Å in the corresponding "concave" structure. It is conceivable that the emerging $O^a_{amide}\cdots O^b_{amide}$ repulsion is responsible for the observed stabilization of the "convex" structure, leading to an endo-coordination of axial water molecules and the corresponding destabilization of the folded structure.

6. ISOMER DISTRIBUTIONS AND REDUCTION POTENTIALS: THE INFLU-ENCE OF THE ENVIRONMENT

Conformational analysis is the pioneering application of molecular mechanics to coordination compounds,[21] and $[Co(dien)_2]^{3+}$ (dien = 1,4,7-triazaheptane (diethylenetriamine), see Table 2) is the most extensively studied system.[3] Generally, entropic terms, other than statistical contributions, and environmental effects (solvation and ion-pairing) have been neglected in molecular mechanics studies of the three isomers of this system. However, based on experimental data, and in particular on those presented in the upper part of Table 2, the neglect of entropic and environmental effects does not seem to be a reasonable approach:

Table 2. Experimentally determined and computed isomer distributions of the $[Co(dien)_2]^{3+}$ system (dien = 1,4,7-triazaheptane)

mer   sym - fac   asym - fac

	mer (%)	sym–fac (%)	asym–fac(%)	Ref.
	experimentally determined			
H_2O, ClO_4^-, Cl^-, Br^-, NO_3^-	63	8	29	
SO_4^{2-} (2M)	37	25	38	24
PO_4^{3-} (0.08M)	20	55	25	
H_2O, CH_3COO^-: (RT)	66	7	27	
(80°)	44	14	42	
H_2O	63	8	29	
MeOH	53	18	29	
DMSO	80	6	14	
Acetone	74	9	17	25
	MM calculations			
	20	40	40	26
	34	34	31	27
	93	2	5	28
MOMEC 92 298 (353) K	66 (57)	2 (3)	32 (40)	29
MOMEC 94 298 (353) K	56 (48)	3 (4)	41 (48)	30
MOMEC 97 298 (313) K	60 (51)	6 (7)	34 (42)	31

(i) the temperature dependence of the conformational equilibrium is underestimated if statistical effects are the only contribution to the entropy term (see last three entries in Table 2); (ii) strong ion-pairing to doubly and triply charged oxo-anions is a well documented effect,[22] and selective ion-pairing, in particular by oxo-anions, is an important feature in the separation of isomers such as those of $[Co(dien)_2]^{3+}$, by ion exchange chromatography and by fractional crystallization;[23] (iii) the differences in experimentally determined isomer distributions in various solvents indicate that the neglect of solvation is not warranted.

The fact that, in many examples, there has been good agreement between observed and predicted isomer distributions is astonishing in view of the data presented in Table 2 and the corresponding comments above. Usually, good agreement is observed when computed distributions are compared to experimental data obtained from aqueous solution with "innocent" anions, i.e. singly charged, hard anions (typically halogenids, perchlorate, etc.; see first entry in Table 2).[3,32,33] Possible reasons for this observation are that (i) ion-pairing involving monoanions, particularly in water, is comparably small;[22] (ii) ion-pairing to small spherical anions is unselective (see also comment on isomer separation above, i.e. efficient separations are typically involving tartrate, phosphate etc.); (iii) the solvation of cationic amine complexes involves amine proton to water hydrogen bonding, and it is conceivable that this is rather unselective. Thus, the environment in aqueous solution involves water molecules and monoanions that are randomly hydrogen-bonded (ion-paired) to the metal-amine cation. This is an environment that is similar to that in a crystal lattice, i.e. comparable to the structures to which the force field has been fitted.[3]

The increasing accuracy in the prediction of the isomer distribution of the $[Co(dien)_2]^{3+}$ system (and other systems reported in the literature[3,33]) is mainly due to two factors (see lower part of Table 2): (i) improved force field parameterization and (ii) thorough analysis of the entire conformational space. The improvement of the force field involved successively the inclusion of 1,3-nonbonded interactions around the metal center (ligand-ligand repulsion, points-on-a-sphere model; 3rd entry in the lower part of Table 2[28]); the thorough analysis of the conformational space (4th entry in the lower part of Table 2; similar force field to that used for the 3rd entry[29]); inclusion of a ligand-field based electronic term in addition to 1,3-nonbonded interactions (5th entry in the lower part of Table 2[30]); modification of the parameterization of the organic backbone of the coordinated ligands (last entry in Table 2[31]).

Reduction potentials of cobalt(III/II) hexaamines[34,35] and of copper(II/I) tetraamines[36] have been computed successfully by a correlation of the strain energy differences between the oxidized and reduced forms with observed reduction potentials. The basis of this approach is that the free energy difference between the oxidized and the reduced form of a redox couple is related to the corresponding reduction potential ($\Delta G^0 = -n \cdot F \cdot E^0$). Assuming that the difference in strain induced by the ligand sphere to the oxidized and the reduced metal cation, and the difference in strain induced by the oxidized and the reduced form of the metal center to the ligand system are of major importance, ΔG^0 may be replaced by the difference in strain energies of the oxidized and the reduced forms of the corresponding metal complexes. A typical plot of calculated strain energy differences versus experimentally determined reduction potentials is given in Fig. 10.[34]

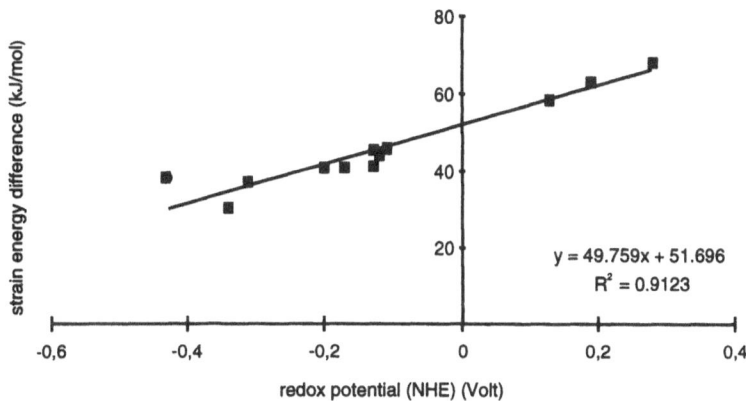

Figure 10. Correlation of calculated strain energy differences and observed reduction potentials for $Co^{III/II}$ hexaamines involving secondary amine donors.[34]

The fact that the slope is smaller than the theoretically expected 96.5 kJ mol^{-1} V^{-1} indicates that electronic effects, the difference in entropy and solvation free energy are of some importance. The observation that the slope is approximately 60 % of that expected theoretically indicates that steric factors are of major importance. The good linearity indicates that the neglected terms (the entropy and the environment) are constant or strongly correlated to the strain energy differences. It has been argued that both the entropic contribution to ΔG^0 and the influence of the environment are correlated to the strain energy differences.[34,36] However, the qualitative arguments will have to be studied in more detail, and this clearly is also of importance for the further evaluation and refinement of the computation of isomer distributions.

Acknowledgment

I am grateful for the generous financial support by the German Science Foundation (DFG), the Fonds of the Chemical Industry (FCI) and the VW-Stiftung, for the excellent work done by my coworkers whose names appear in the references and for the help by Karin Stelzer in preparing this manuscript.

REFERENCES

1. J.-M. Lehn. "Supramolecular Chemistry – Concepts and Perspectives", VCH, Weinheim (1995).
2. J.E. Huheey, "Inorganic Chemistry – Principles of Structure and Reactivity", Harper, New York, 2nd edn. (1978).
3. (a) P. Comba and T. W. Hambley, "Molecular Modeling of Inorganic Compounds". VCH, Weinheim (1995); (b) P. Comba, *in*: "Fundamental Principles of Molecular Modeling", W. Gans, A. Amann and J. C. A. Boeyens, eds., p. 169, Plenum Press, New York, (1996); (c) P. Comba, *in* "Molecular Modeling and Dynamics of Bioinorganic Compounds", Kluwer Academic Publishers, Dordrecht, Boston, London (1997).
4. (a) J. Sabolovic and K. Rasmussen, *Inorg. Chem.* 34:1221 (1995); (b) L. Glasser, *in*; "Fundamental Principles of Molecular Modeling", W. Gans, A. Amann and J.C.A. Boeyens, eds., p. 199, Plenum Press, New York (1995).
5. (a) B.J. Hathaway and P.G. Hodgson, *Inorg. Nucl. Chem.* 35:4071 (1973); (b) J. Gazo, I.B. Bersuker, J. Garaj, J. Kabesova, H. Langfelderova, M. Melni, M. Serator and F. Valack, *Coord. Chem. Rev.* 19:253 (1976).
6. H. Oshio, *Inorg. Chem.* 32:4123 (1993).
7. T.W. Hambley, *J. Chem. Soc., Dalton Trans.* p. 565 (1986).
8. P. Comba, A. Peters and H. Pritzkow, Structure and properties of a copper(II) tetraamine with an extremely short axial bond, publication in preparation.
9. P. Comba, P. Hilfenhaus and B. Nuber, *Helv. Chim. Acta* 80:1831 (1997).
10. P. Comba, T. W. Hambley, M. A. Hitchman and H. Stratemeier, *Inorg. Chem.* 34:3903 (1995).
11. M.A. Hitchman, *Transition Met. Chem.* 9:1 (1985).
12. (a) C.E. Schäfer and C.K. Jørgensen, *Mol. Phys.* 9:401 (1965); (b) E. Larsen and G.N. LaMar *J. Chem. Educ.* 51:633 (1974).
13. M. Saburi, K. Miyamura, M. Morita, Y. Mizoguchi, S. Yoshikawa, T. Tsuboyama, T. Sakurai and K. Tsuboyama, *Bull. Chem. Soc. Jpn.* 60:141 (1987).
14. P. Comba, T. W. Hambley, G. A. Lawrance, L. L. Martin, P. Renold and K. Vrnagy, *J. Chem. Soc., Dalton Trans.* p. 277 (1991).
15. J. Balla, P. V. Bernhardt, P. Buglyo, P. Comba, T.W. Hambley, R. Schmidlin, S. Stebler and K. Várnagy, *J. Chem. Soc., Dalton Trans.* p. 1143 (1993).
16. P.V. Bernhardt, P. Comba, T.W. Hambley, G.A. Lawrance and K. Várnagy, *J. Chem. Soc., Dalton Trans.* p. 355 (1992).
17. P. Comba, N. F. Curtis, G. A. Lawrance, M. A. O'Leary, B. W. Skelton and A H. White, *J. Chem. Soc., Dalton Trans.* p. 2145 (1988).
18. (a) W.L. Driessen and W.L. Groeneveld, *Rec. Trav. Chim.* 88:491 (1969); (b) W.L. Driessen and W.L. Groeneveld, *Rec. Trav. Chim.* 88:620 (1969); (c) W.L. Driessen and W.L. Groeneveld, *Rec. Trav. Chim.* 89:1271 (1970),.
19. P. Comba, S. P. Gavrish, R. W. Hay, P. Hilfenhaus, Y. D. Lampeka, P. Lightfoot and A. Peters, Analysis and interpretation of significant structural differences of dinuclear complexes (M = Ni(II), Cu(II)) of a bismacrocyclic ligand, publication in preparation.
20. (a) P. V. Bernhardt, P. Comba, T. W. Hambley, S. S. Massoud and S. Stebler, *Inorg. Chem.* 31:2644 (1992); (b) P. Comba, *Comm. Inorg. Chem.* 16:133 (1994); (c) P. Comba and P. Hilfenhaus, *J. Chem. Soc., Dalton Trans.* p. 3269 (1995); (d) P. Comba, T.W. Hambley, P. Hilfenhaus and D.T. Richens, *J. Chem. Soc., Dalton Trans.* p. 533 (1996).
21. (a) J.-P. Mathieu, *Ann. Phys.* 19:335 (1944); (b) E.J. Corey and J.C. Bailar Jr., *J. Am. Chem. Soc.* 81:2620 (1959).
22. W. G. Jackson, M. L. Hookey, M. L. Randall, P. Comba and A. M. Sargeson, *Inorg. Chem.* 23:2473 (1984).
23. G.H. Searle, *Aust. J. Chem.* 30:2525 (1977).
24. F.R. Keene and G.H. Searle, *Inorg. Chem.* 13:2173 (1974).
25. A.M. Bond, F.R. Keene, N.W. Rumble, G.H. Searle and M.R. Snow, *Inorg. Chem.* 17:2847 (1978).

26. M. Dwyer and G.H. Searle, *J. Chem. Soc., Chem. Comm.* p. 726 (1972).

27. Y. Yoshikawa, *Bull. Chem. Soc. Jpn.* 49:159 (1976).

28. A.M. Bond, T.W. Hambley and M.R. Snow, *Inorg. Chem.* 24:1920 (1985).

29. P. V. Bernhardt and P. Comba, *Inorg. Chem.* 31:2638 (1992).

30. P. Comba, T. W. Hambley and M. Ströhle, *Helv. Chim. Acta,* 78:2042 (1995).

31. J. E. Bol, C. Buning, P. Comba, J. Reedijk and M. Ströhle, *J. Comput. Chem.* (1997), in press.

32. P. Comba, T. W. Hambley and L. Zipper, *Helv. Chim. Acta* 71:1875 (1988).

33. P. Comba, *Coord. Chem. Rev.* 123:1 (1993).

34. P. Comba and A. F. Sickmüller, *Inorg. Chem.* 36:4500 (1997).

35. P. Comba and A. F. Sickmüller, *Angew. Chem.* 109:2089 (1997); *Angew. Chem. Int. Ed. Engl.* 36:2006 (1997).

36. P. Comba and H. Jakob, *Helv. Chim. Acta* 80:1983 (1997).

RELATIONSHIPS BETWEEN EXPERIMENT AND THEORY IN THE STUDY OF INTERMOLECULAR INTERACTIONS

Frank H. Allen

Cambridge Crystallographic Data Centre
12 Union Road
Cambridge CB2 1EZ
England

1. INTRODUCTION

Systematic knowledge of the types, geometries and strengths of a wide variety of non-covalent interactions is crucial in many areas, for example supramolecular chemistry, crystal engineering, protein-ligand docking and rational drug design. A crystal structure is the archetypal supermolecule and every structure provides direct experimental observations of the types and geometries of those intermolecular interactions that are responsible for molecular aggregation in the crystalline state. Geometrical data derived from, for example, the Cambridge Structural Database (CSD) of small molecule crystal structures,[1] can provide systematic knowledge of frequencies of formation, dimensions and directional preferences exhibited by specific non-covalent bonds.

However, intercomparisons of statistical information, e.g. a comparison of hydrogen bond lengths in different donor–acceptor systems, or comparisons of the frequencies of formation of dissimilar interactions, can only indicate an approximate ordering of non-covalent bond energies. These energies are vital information if we are to engineer supramolecular units which have predictable intermolecular relationships, and it is therefore important to supplement these experimental observations with high-level theoretical calculations of the interaction energies of selected model dimers.

This Chapter describes how the experimental data stored in the Cambridge Structural Database can be searched and analysed to obtain systematic information about intermolecular interactions. It then goes on to show how this information can be used to guide high level *ab initio*-based molecular orbital calculations on model dimers, using intermolecular perturbation theory,[2] so as to augment and explain some of the features, and apparent inconsistencies, in the experimental observations. Finally, we describe how rapid access to experimental and computational information about more than 6,000 non-covalent interactions is now provided via an interactive knowledge-based library - the IsoStar library[3] – the first knowledge-based product of the Cambridge Crystallographic Data Centre.

2. METHODS

Database Analyses

The CSD currently contains bibliographic, chemical connectivity and three-dimensional coordinate data for more than 170,000 organic, organometallic and metal complex crystal structures. The CSD System also comprises an interactive suite of programs for search and retrieval (Quest3D), structure visualisation (Pluto), and for the statistical analysis and display of derived geometrical information (Vista). The overall operation of the CSD System is fully described elsewhere;[1] the most important functions that are relevant to the current topic are:

(a) The ability to search (Quest3D) the extended crystal structure for intermolecular interactions that are specified in terms of the atoms or functional groups involved in the interaction, together with geometrical constraints (distances, angles, etc.) that define the presence or absence of the interaction

(b) Once located, Quest3D will then calculate a wide variety of geometrical descriptors for the interaction, over and above those used in its search definition. A typical search fragment, for hydrogen bonds to thiocarbonyl groups,[4] is shown in Figure 1. Here, the search was carried out using $d(SH) < 2.9$ Å, and $\rho(H) > 90°$. However, to describe the interaction more completely, Quest3D was also required to calculate the directionality parameters ϑ and ϕ (Figure 1) that describe the direction of approach of the donor-H to the plane that contains the S-lone pairs.

(c) The selected geometrical descriptors for each hit located by Quest3D are output to a file for use by Vista. This program can display histograms, scattergrams, polar plots, etc. of the geometrical data, and also perform a variety of statistical functions: simple descriptive statistics, regression analyses, principal component analyses, etc. Some of the Figures in this chapter were generated using the Vista program.

Molecular Orbital Calculations

Intermolecular perturbation theory (IMPT),[2] as implemented within the CADPAC6.0 program package,[5] was used to quantify intermolecular interaction energies. Simple model monomer molecules were constructed and their geometries were optimised using 6-31G* or 6-31G** basis sets as appropriate to the systems under study. Interacting dimers were then constructed using geometrical relationships within the dimer that were indicated by the relevant database analysis; intermolecular distances involving non-interacting atoms of the dimer were kept as long as possible. IMPT calculations are cpu-intensive, and cpu-times rise dramatically as the number of atoms in the dimer model increases. For this reason, dimer components were kept as small as possible, consistent with being fully representative of the non-covalent interaction under study. Further, IMPT is an inappropriate technique for the systematic sampling of the potential energy hypersurface, hence the use of preferred geometries indicated by the database analyses to establish the initial mutual orientations of dimer components. Essentially, we regard the crystallographically preferred interaction geometries as representing one or more minima in the hypersurface. From this starting point (or points), IMPT calculations were carried out in which each of the geometrical parameters was allowed to vary away from its crystallographically preferred value(s), so as to scan the most likely minimum energy areas and establish some estimate of the interaction energies in those areas. Application of the IMPT method within CADPAC is fully described elsewhere[6] and only a brief summary is given here.

An important feature of the IMPT method is that it calculates separate interaction energy terms which have distinct physical significance. Further, the sum of the significant interaction energy terms yields a total IMPT energy which is free of basis set superposition error.[7] At first order, these separate terms are (a) E_{es}: the (attractive or repulsive) electrostatic energy that describes the classical Coulombic interaction, and (b) E_{er}: the exchange–repulsion term, the sum of an attractive part due to the exchange of electrons of parallel spin, and a repulsive part as a result of the Pauli exclusion principle which prevents electrons with parallel spins occupying the same region in space. At second order, the IMPT gives (c) E_{pol}: the polarisation or induction energy, (d) E_{ct}: the charge transfer energy, and (e) E_{disp}: the dispersion energy term. We note that the dispersion energy is largely dependent on the size of the basis set used. Thus, even the 6-31G* basis set is too small to describe accurately the dynamic polarisability that underlies the dispersion energy, and this results in an underestimation of E_{disp}. However,[6,7] dispersion effects are only weakly orientation dependent and their underestimation does not affect the validity of the general conclusions derived in this work.

3. APPLICATIONS

Hydrogen Bonding to Thiocarbonyl-S Acceptors

Given the minimal electronegativity difference between C and S, it may be considered surprising that H-bonds to thiocarbonyl acceptors form at all. However, the C=S bond is highly polarisable, and shows a very broad interatomic distance range[4] from 1.58 to 1.76 Å. Recently,[4] a systematic survey of crystallographic data has shown that:

(a) The frequency of H-bond formation to C=S acceptors increases systematically as $d(CS)$ increases in length: 23.5 % of S accept H-bonds at $d(CS) = 1.63$Å, while 74.2 % accept H-bonds at $d(CS) = 1.73$ Å. It was shown[4] that the required δ-negativity at S, and the lengthening of $d(CS)$, is induced by the presence of electron-rich centres, e.g. N atoms, at R_1, R_2 in Figure 1.

(b) Analogous to H-bonds to >C=O acceptors, H-bonds to >C=S acceptors tend to form in the >C=S (lone-pair) plane: there is a clear preference for ϑ-angles (Figure 1) close to zero in the distribution of experimental values.

(c) Within the >C=O, >C=S planes, Figure 2 shows that there is a preference for donor-H to approach O at ϕ-angles (Figure 1) in the narrow range 120–130° – a clear indication of lone-pair directionality in the >C=O···H system. This directionality is also clearly observed in the ϕ-distribution for >C=S···H (Figure 2), but at significantly lower angles, of ca. 100°, in the case of S.

(d) H-bonds to >C=S are significantly (0.25 Å) longer (hence weaker) than those to >C=O, as shown by a comparison of mean lengths for S···H−N and S···H−O hydrogen bonds with their O-acceptor counterparts, in which the difference in van der Waals radii of S and O was taken into account.

We have now used the IMPT method[8] to obtain further comparative insight into the >C=S···H−D and >C=O···H−D systems. Preliminary results of calculations at the 6-31G* level are shown in Table 1. Model dimers used methanol to provide donor O−H, with the O−H vector in the >C=S,O plane, and the H−O−C(methyl) plane perpendicular to it. Optimum S···H and O···H separations, originally taken as 2.5 and 1.9 Å from the crystallographic analysis, were refined to 2.55 and 1.82 Å on the basis of IMPT calculations for a range of values of $d(SH)$ and $d(OH)$. IMPT calculations also confirmed the preference for in-plane H-approach at $\vartheta = 0°$.

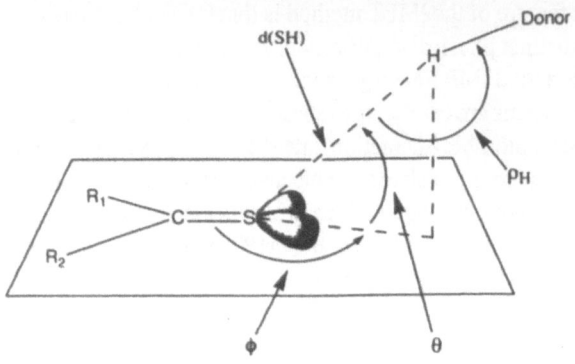

Figure 1. Search fragment and geometrical descriptiors for the $>C=S\cdots H$-donor system.

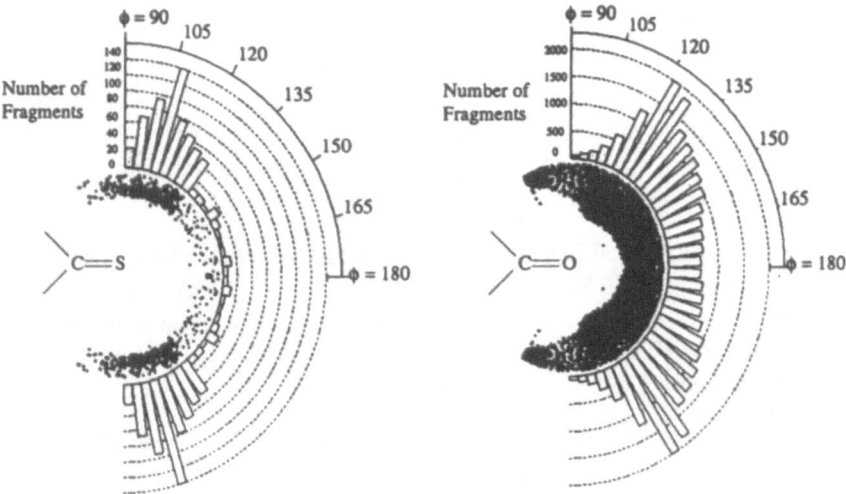

Figure 2. Polar histograms of ϕ (Figure 1) from database analyses of $>C=S\cdots H$ systems (left) and $>C=O\cdots H$ systems (right).

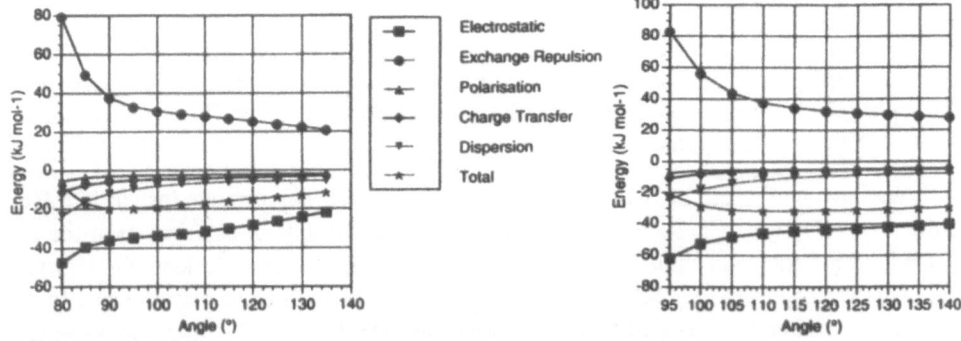

Figure 3. IMPT energies for values of the angle ϕ in the range 80–140° for the thiourea–methanol model (left) and for the urea–methanol model (right), using a fixed ϑ of zero and a fixed $d(SH)$ or $d(OH)$ of 2.55 or 1.82 Å as appropriate.

Table 1. IMPT results for $(R_1)(R_2)C=S$-methanol and
$(R_1)(R_2)C=O$-methanol dimers

R_1	CH$_3$	CH$_3$	NH$_2$
R_2	CH$_3$	NH$_2$	NH$_2$
		$(R_1)(R_2)C=S$-methanol dimer	
$d(C=S)$ (Å)	1.620	1.644	1.661
E_t (kJ mol^{-1})	-12.7	-18.5	-20.1
ϕ (deg.)	100	95	93
		$(R_1)(R_2)C=O$-methanol dimer	
$d(C=O)$ (Å)	1.225	1.227	1.227
E_t (kJ mol^{-1})	-25.0	-30.4	-32.0
ϕ (deg.)	125	110	110

The computational data of Table 1 agree well with the experimental results derived by database analysis. In particular, H-bonds to the S-acceptors are very significantly weaker (E_t) than those to their O-counterparts. H-bonds to dimethylthione are weak, being similar in strength to C$-$H$\cdots$O bonds, and it is only with the introduction of one, or preferably two, N-substituents that $>$C$=$S$\cdots$H bonds can be classified as being of medium strength. In light of Figure 2, the optimum values of ϕ given by the IMPT calculations are of special interest. The IMPT energy profiles for the thiourea–methanol and urea–methanol dimers are compared in Figure 3. Both plots show that the overall interaction energy (E_t) results from a balance between attractive electrostatic interactions and the exchange repulsion terms. In both cases, there is a broad minimum in the interaction energy profile, with the most attractive interaction energies occurring between 90 and 100° for thiourea, but above ca. 110° for urea: an angular difference that is in clear agreement with the results of Figure 2.

Dipole–Dipole Interactions Between Carbonyl Groups

A hydrogen bond is, perhaps, the archetypal dipole–dipole interaction. However, many such dipolar interactions not mediated by δ-positive H have been observed to have a significant influence on molecular packing in crystals. One such is the interaction between pairs of carbonyl dipoles $C(\delta+)=O(\delta-)$. Despite the presence of appropriate donor-H atoms, molecules of alloxan are observed[9] to aggregate in the crystal through antiparallel interactions between C$=$O dipoles. Others[10-13] have remarked on the prevalence and likely energetic importance of C$=$O$\cdots$C$=$O interactions in crystal packing, and the importance of this interaction in stabilizing protein secondary structure, particularly helices, β-sheets and β-strands, has recently been highlighted.[14,15]

In a recent systematic CSD study,[16] 9049 $>$C$=$O groups were retrieved and 1328 (15 %) of them were found to have close contacts (within van der Waals radii sums $+0.2$ Å) with other $>$C$=$O groups. By comparison, 41 % of $>$C$=$O groups from crystal structures that also contain O$-$H or N$-$H donors actually form hydrogen bonds. From these simple statistics, we can infer that dipolar interactions between C$=$O groups are important, but are much weaker than $>$C$=$O$\cdots$H hydrogen bonds. The complete analysis of CSD data showed that three distinct interaction motifs occur in crystal structures (Figure 4): (a) an antiparallel motif which is slightly sheared and involves two close C$\cdots$O interactions, and (b,c) a perpendicular motif and a sheared parallel motif, both of which involve a single short C$\cdots$O interaction. Motif (a) is clearly preferred by a ratio of ca. 5:1 over either of the other motifs.

115

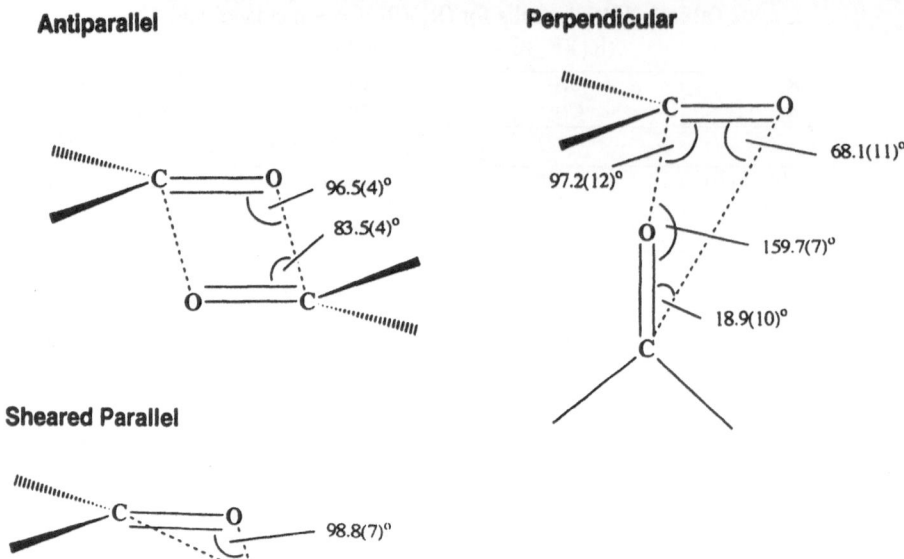

Figure 4. The three most common carbonyl dipole interaction motifs observed in the CSD.

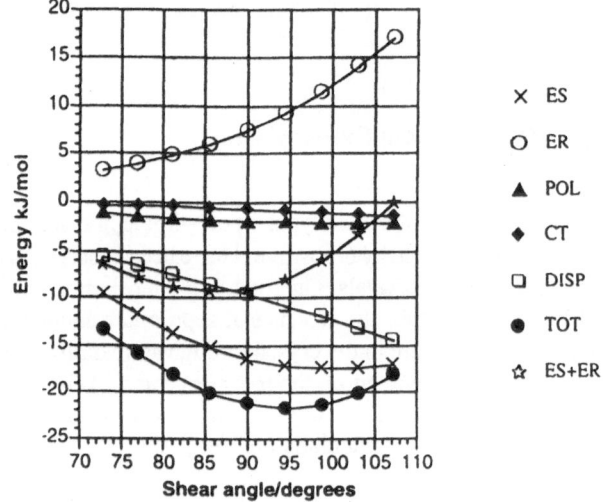

Figure 5. IMPT energy profiles for variation of the shear angle A1 (see text).

The IMPT method has been used[16] to quantify the energetic importance of the motifs of Figure 4, using a simple bis - propanone dimer model. For the antiparallel motif, the four C and O atoms were kept coplanar (as clearly indicated by the crystallographic data), and the interaction distance $d(CO)$ and the angle C=O···C (denoted A1) were varied. For a perfectly rectangular motif, the minimum interaction energy was located at $d(CO) = 3.02$ Å, a value towards the lower end of the distance distribution obtained from the CSD. However (Figure 5), as the distance is increased, IMPT indicates that a slightly sheared motif, having a shear angle A1 of ca. 95°, is preferred, and without significant loss of attractive energy: in either case, a minimum in the E_t profile close to -22.0 kJ mol^{-1} is obtained. This value is comparable to the IMPT results for the medium strength $>C=S\cdots H$ system (Table 1), but the dipolar interaction is considerably less attractive than the strong $>C=O\cdots H$ system. Again, Figure 5 shows that this energy minimum reflects a balance between exchange repulsion and attractive electrostatic interactions. IMPT calculations for the perpendicular motif yield an overall E_t of -8 kJ mol^{-1}, a value which is unsurprising since only a single close $C\cdots O$ interaction is involved in this motif, and in the sheared parallel motif (c) of Figure 4. All of the IMPT results for this system are clearly in agreement with the energetic inferences that can be made from the experimental information retrieved from the CSD.

Other Combined Crystallographic and Molecular Orbital Studies

Interactions of halogen atoms with oxygen or nitrogen have been a subject of study by a number of authors (see ref. 6). The preference for formation of, e.g., short $C-Cl\cdots O=C$ contacts that are collinear with the $C-Cl$ bond has been ascribed to a polar flattening effect at Cl: a smaller effective van der Waals radius along the bond extension, by comparison with the effective radius perpendicular to the bond. Recently[6] a thorough study involving database analysis and IMPT calculations showed that (a) a much larger than random number of O and N atoms penetrate the halogen sphere in the linear (C–Hal) direction, the proportion increasing in the order Cl $<$ Br $<$ I, but (b) this behaviour was not reproduced by approaching H atoms. The IMPT calculations, using a chloro-cyanoacetylene dimer model, reveal the attractive nature of the $Cl\cdots N$ interaction and give an interaction energy of ca. -10 kJ mol^{-1}, principally due to electrostatic attraction, although polarisation, charge-transfer and dispersion terms are also important. The directionality of the interaction is thus explained by an anisotropic electron distribution around the halogen, leading to a decreased repulsive wall and increased electrostatic attraction for electronegative atoms in the observed forward direction.

The attractive energies of these halogen$\cdots$oxygen interactions provide a rationale[6] for the different packing arrangements adopted by 2,5-dichloro- and 2,5-dibromo-1,4-benzoquinone,[17,18] which are characterised by short $Cl\cdots O$ and $Br\cdots O$ contacts respectively, and that of 2,5-dimethyl-1,4-benzoquinone,[19] where $C-H\cdots O$ interactions of similar energy now determine the molecular arrangement. The importance of $Hal\cdots O$ interactions is also illustrated by the use of the symmetrical bifurcated nitro-$O\cdots$iodo supramolecular synthon[20] in engineering the structures[21,22] illustrated in Figure 6. A combined database and IMPT study of nitro-$O\cdots$halogen systems[23] shows that 45 % of these interactions are less than the sum of van der Waals radii, with an increasing preference for the symmetrical bifurcated motif in the order Cl $<$ Br $<$ I. IMPT calculations at the 6-31G* level were only possible for nitro-$O\cdots$Cl systems and yielded an attractive energy of -6 kJ mol^{-1}. However, lower level calculations for $Br\cdots O$ indicate that this interaction is more attractive than for Cl, and this result, together with the relative shortening of the nitro-$O\cdots$I interactions in the database analysis, would imply an even stronger $I\cdots O$ interaction. A similar combined study of H-bond formation to nitro-O acceptors has also been carried out.[24]

Molecular orbital calculations are also valuable in explaining apparent anomalies in database observations.[25] Thus, it is well known that carbonyl and ether oxygens are potent acceptors of donor-H atoms. However, in crystallographic studies of E-esters a very low density of H-bonds is found at the etheric ester-O atom, while the acceptor ability of the terminal $C=O$ atom is as expected. IMPT calculations show that this is essentially a competition effect: the ester $-O-$ atom remains a reasonably strong acceptor with $O\cdots H$ in the computational model having an energy of -20.4 kJ mol^{-1}, but the strength of the $C=O\cdots H$ bonds is significantly larger, at -27.2 kJ mol^{-1}, thus accounting for the preferential formation of H-bonds at this acceptor.

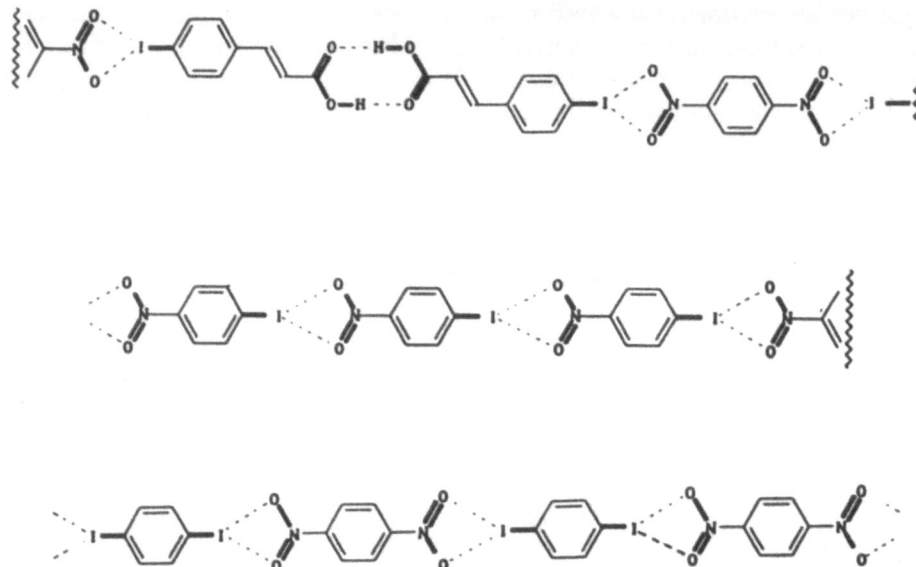

Figure 6. Crystal structures[21,22] engineered using the iodo–nitro supramolecular synthon.

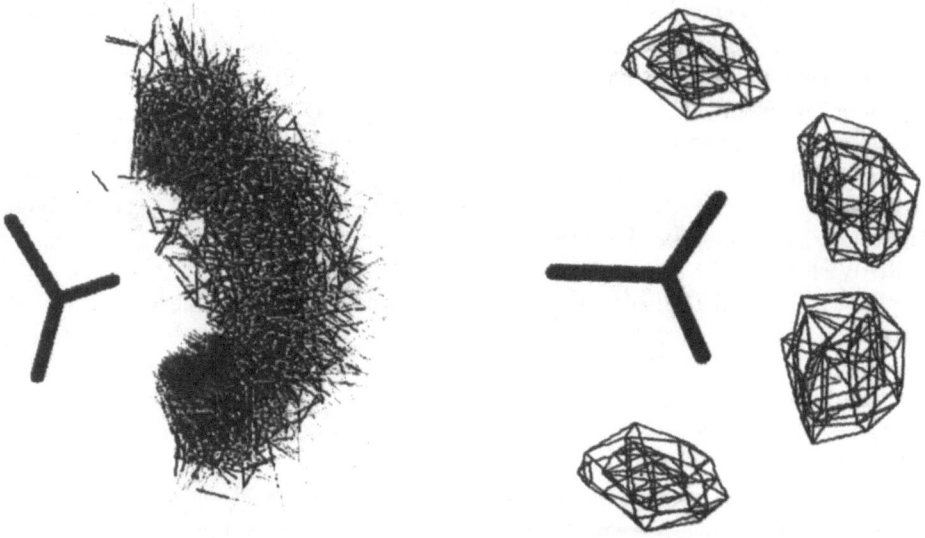

Figure 7. Examples of information presentation from the IsoStar library of intermolecular interactions: the distribution of O−H donors around a carboxylate oxygen acceptor is shown as an unsymmetrised vector plot (left), and as a symmetrised contour plot (right).

4. A KNOWLEDGE-BASED LIBRARY OF INTERMOLECULAR INTERACTIONS

Given the importance of geometrical and other knowledge concerning intermolecular interactions, and in view of the often time-consuming nature of the database searches, subsequent data analyses and, especially, the cpu-intensity of the molecular orbital calculations, there is a need to collect the required information into a more accessible form for the end user. Over the past three years, and building on its research experiences, the CCDC has been developing a library – denoted as IsoStar – containing systematic information about intermolecular interactions.[3] Much of this information is derived from the raw crystallographic data held in the CSD, but information from the Protein Data Bank (PDB), and from extensive application of IMPT and other theoretical chemistry calculations, is also included. PDB information is used from those protein–ligand complexes that have a resolution of 2.5 Å or better.

Central to the IsoStar library design is an extensive set of scatterplots (see e.g. Figure 7) which show superimposed views of the disposition of a given *contact group* (here O−H donors) around a specified *central group* (here a carboxylate group). Scatterplots incorporate contacts out to van der Waals radii sums +0.4 Å, and can be rotated, zoomed, etc., in an interactive manner. Choices of central and contact group combinations, and of the data source (CSD, PDB, MO), are made from a Web-browser interface, and a number of IsoStar windows then permit interactive control over, inter alia, (a) the non-covalent distance limit to be applied to the current scatterplot, (b) the display of geometry for a specific interaction, (c) the display of a specific interaction within its original structural context (via hyperlinks to the CSD or PDB), and (d) facilities that present the scatterplots as contoured surfaces, as depicted in Figure 7.

Version 1.0 of IsoStar was released as an additional module of the CSD System in October 1997. It contains information about 277 central groups and 28 contact groups – giving a total of 7756 possible scatterplots from each of the CSD and the PDB. Version 1.0 of IsoStar contains a total of 6683 scatterplots: 5296 derived from the CSD and 1387 from the PDB. In addition, 367 model systems have been studied using MO methods, and 867 potential energy minima are recorded in IsoStar.

The IsoStar library will be maintained on a regular basis as the content of the CSD and PDB increases. It is also hoped to further develop IsoStar through inclusion of some or all of: statistical data relating to distributions of geometrical parameters, information about common motifs formed with the involvement of non-covalent bonds, information on preferred metal coordination modes, etc.

5. CONCLUSION

This paper has illustrated the value of experimental observations of intermolecular interactions, and how some of the systematic knowledge derived from data sources such as the CSD or PDB can be interpreted and enhanced through the careful use of high-level *ab initio* molecular orbital calculations. A number of basic research projects have indicated the value of this combination of techniques, and have contributed to the inception of a major new development – the IsoStar knowledge-based library of intermolecular interactions. This computerised library not only brings structural knowledge to the computational or synthetic chemist in a readily accessible form, it also provides a basis for new software approaches to problem solving in the areas of structural chemistry and rational drug design. One such development is that of the GOLD program[26] for protein–ligand docking, a program that uses genetic algorithms, and already incorporates conformational information and knowledge of non-covalent interactions derived from the CSD in deriving its fitness functions.

REFERENCES

1. F.H. Allen, J.E. Davies, J.J. Galloy, O. Johnson, O. Kennard, C.F. Macrae, E.M. Mitchell, G.F. Mitchell, J.M. Smith and D.G. Watson, *J. Chem. Inf. Comput. Sci.* 31:187 (1991).
2. I.C Hayes and A.J. Stone, *J. Mol. Phys.* 53:83 (1984).
3. I.J. Bruno, J.C. Cole, J.P.M. Lommerse, R.S. Rowland, R. Taylor and M.L. Verdonk, *J. Computer-Aided Mol. Design.*, in press.
4. F.H. Allen, C.M. Bird, R.S. Rowland and P.R. Raithby, *Acta Cryst.* B53:696 (1997).
5. R.D. Amos, (1996). CADPAC6.0. The Cambridge Analytical Derivatives Package. Issue 6.0. A suite of quantum chemistry programs. Department of Chemistry, University of Cambridge, Lensfield Road, Cambridge CB2 1EW, England.
6. J.P.M. Lommerse, A.J. Stone, R. Taylor and F.H. Allen, *J. Am. Chem. Soc.* 118:3108 (1996).
7. A.J. Stone, *Chem. Phys. Lett.* 211:401 (1993).
8. F.H. Allen, J.A.K. Howard and H. Amer, *Acta Cryst.* Section B, submitted.
9. W. Bolton, *Acta Cryst.* 18:5 (1965).
10. J. Bernstein, M.D. Cohen and L. Leiserowitz, *in*: "The Chemistry of Quinonoid Compounds", S.Patai, ed., 83-105, Wiley, London (1974).
11. H.-B. Bürgi, J.D. Dunitz and E. Shefter, *Acta Cryst.* B30:1517 (1974).
12. A. Gavezzotti, *J. Phys. Chem.* 94:4318 (1990).
13. R. Taylor, A. Mullaley and G.W. Mullier, *Pesticide Sci.* 29:197 (1990).
14. P.H. Maccallum, R. Poet and E.J. Milner-White, *J. Mol. Biol.* 248:361 (1995).
15. P.H. Maccallum, R. Poet and E.J. Milner-White, *J. Mol. Biol.* 248:374 (1995).
16. F.H. Allen, C.A. Baalham, J.P.M. Lommerse and P.R. Raithby, *Acta Cryst.* Section B, in press.
17. B. Rees, *Acta Cryst.* B26:1304 (1970).
18. B. Rees, R. Haser and R. Weiss, *Bull. Soc. Chim. Fr.* p. 2568 (1966).
19. F.L. Hirshfeld and D. Rabinovich, *Acta Cryst.* 23:989 (1967).
20. G.R. Desiraju, *Angew. Chem. Int. Ed. Engl.* 34:2311 (1995).
21. F.H. Allen, B.S. Goud, V.J. Hoy, J.A.K. Howard and G.R. Desiraju, *Chem. Commun.* p. 2729 (1994).
22. V.R. Thalladi, B.S. Goud, V.J. Hoy, F.H. Allen, J.A.K. Howard and G.R. Desiraju, *Chem. Commun.* p. 401 (1996).
23. F.H. Allen, J.P.M. Lommerse, V.J. Hoy, J.A.K. Howard and G.R.Desiraju, *Acta Cryst.* B53:1006 (1997).
24. F.H. Allen, C.A. Baalham, J.P.M. Lommerse, P.R. Raithby and E. Sparr, *Acta Cryst.* B53:1017 (1997).
25. J.P.M. Lommerse, S.L. Price and R. Taylor, *J. Comput. Chem.* 18:757 (1997).
26. G. Jones, P. Willett, R.C. Glen, A.R. Leach and R. Taylor, *J. Mol. Biol.* 267:727 (1997).

STUDY OF INTERMOLECULAR INTERACTIONS USING CRYSTAL STRUCTURE DATABASE AS REFERENCE: A PRELIMINARY REPORT ON THE ADJUSTMENT OF VAN DER WAALS CONSTANTS

Eiji Ōsawa,[a]* Hitoshi Gotō,[b] Takako Sugiki,[a] and Keisuke Imai[c]

[a]Department of Knowledge-based Information Engineering Faculty of
Engineering Toyohashi University of Technology
Tempaku-cho Toyohashi, 441, Japan
[b]Institute for Chemical Reaction Science Tohoku University
Katahara, 2-1-1, Aoba-ku Sendai, 980-77, Japan
[c]Exploratory Research Laboratories Fujisawa Pharmaceutical Co., Ltd.
5-2-3 Tokodai, Tsukuba Ibaraki 300-26, Japan

1. INTRODUCTION

The Cambridge Structure Database (CSD) of X-ray crystal structures[1] provides an immense treasurehouse of non-covalent intermolecular interactions. One way of making wise use of the unprecedented situation that more than 120,000 high-quality data on the crystal packing of molecular solids can be readily retrieved at our fingertips will be to re-formulate potential functions of van der Waals and other weak intermolecular interactions adopted in the empirical molecular mechanics schemes by using CSD as the reference. In spite of the high research activities aimed at improving potential functions for molecular dynamics, Monte Carlo calculations, and other atomistic simulations of chemical phenomena, the crystal structure approach has, to our knowledge, never been adopted. The use of high-level ab initio computational results as the standard has been eagerly pursued for the past decade,[2] and proved effective in constructing force fields for dynamic phenomena such as chemical reactions.[3] However, a large enough reference dataset with high enough accuracy is yet to appear. It should be recalled that molecular orbital calculations are by no means the most suitable method for describing weak intermolecular interactions.

As a part of our long-standing project towards the goal of "predicting crystal structures from molecular formulae",[4] we recently examined the crystal structure approach. In retrospect, we first thought that a little re-adjustment of van der Waals constants in a molecular mechanics (MM) force field would lead to improved performance in reproducing crystal structures. However, this anticipation proved premature: the van der Waals constants changed to a large extent and the results forced us to ponder on the potential defect in the MM scheme used. We describe below a preliminary account of the story.

---

*correspondence to osawa@cochem2.tutkie.tut.ac.jp

2. OPTIMIZING CRYSTAL STRUCTURES

A brief mention will be given here on the progress in our peripheral program KESSHOU (meaning crystal),[5] which is designed to simultaneously optimize crystal and molecular structures by growing an ultrafine spherical crystal under appropriate MM scheme without using periodic boundary condition.[6] KESSHOU is implemented in a general molecular mechanics framework program BIGSTRN3,[7,8] and is now upgraded to include polar interactions (dipole/dipole and charge/charge) and to accommodate more than 60,000 atoms.[†]

By using the enhanced version of KESSHOU in combination with MM2[9] and MM3,[10] crystal structures of p-dicyanobenzene, p-diisocyanobenzene and C_{60} were optimized. The observed cell constants were rather well reproduced, but the computed cohesive energies did not agree well with the expected.[11] We thought that the unsatisfactory results reflect the inadequacy of the van der Waals constants of the MM schemes used. Hence the present work was begun.

3. OPTIMIZING VAN DER WAALS CONSTANTS

Our purpose here is two-fold: to develop appropriate method for optimizing constants of molecular mechanics potential functions using crystal structures as reference and then to apply the method to the intermolecular van der Waals potential while keeping other intramolecular interaction potentials intact.

Reference Dataset

Every work on the potential functions for molecular interactions naturally starts with hydrocarbons. We have so far created two kinds of reference dataset. One consists of a total of 208 crystal structures of hydrocarbons selected from more than 1,000 hydrocarbon crystal data registered in the 1995 version of CSD. The selection criteria are (1) R value smaller than 10 %, (2) small to medium molecular size, (3) wide distribution of various space groups, and (4) to avoid like molecules. This dataset was called ALL208. As shown below, however, the first dataset turned out too large to achieve acceptable fitness.

The second dataset was chosen from ALL208 by using two new criteria: (1) R value smaller than 5.0 %, and (2) the initial gradient of intermolecular interaction energy with respect to cell constants not exceeding 50 kcal mol^{-1} Å^{-1}. In this way a smaller dataset consisting of 39 crystal structures were set up and named ALL039.[‡]

Force Field

Allinger's molecular mechanics scheme MM3[10] was chosen, simply because it is one of the most extensively used among chemists in the academic circle, and also because of our own familiarity with the program.[12]

Intermolecular van der Waals interactions were evaluated by using the same exp-6 type function as used in MM3:

$$E_{\text{inter}} = \sum_{i,j} A_{ij} \exp\left(\frac{R_{ij}}{B_{ij}}\right) + \frac{C_{ij}}{R_{ij}^6} \tag{1}$$

[†]The upgrating was carried out by Dr. Petko M. Ivanov during his stay in the author's laboratory as a NEDO research fellow in 1996-1997.

[‡]ADAMAN08, ANTCEN13, BAPBID10, BITTED10, CIWRAB, CLOPNA, CLOPNC, CUJTIK10, DABPIF, DESSAV, DEZNAX, DIHDON, DILCIK, DIWDES, FEVCAK10, FEWWEJ, FILBIL, FIXWAK, GAYTAB, GOJDAK, JAKROC, KEDJOS, KEGJUB, KETREG, KUKDID, LAXXUD, LEFVAT, NAPHTA12, OC-METD10, PASZAK, PELVIL, SAPTIM, VICPAY, YADCUB, YESYOK, YOKJIR, YUBTIY, ZEXREZ, ZZZMKS01

where R_{ij} is the distance between atom i in the central molecule and atom j in a surrounding molecule. Central and surrounding molecules have special meaning in KESSHOU program: only one molecule in the central lattice of the spherical crystal is geometry-optimized while all other surrounding molecules are kept at the geometry from the previous iteration.

Although the van der Waals constants are fixed for each atom type in the MM3 scheme, we give here new atom types to the same hybridization if the environments are different, and the following atom types are considered for hydrocarbons (Table 1). The offset of C–H bond distance conventionally used in MM3 was not considered here.

Table 1. Definitions of Atom Types

Element	Atom type	
C	1	sp^3 carbon atom
C	2	sp^2 carbon atom
C	4	sp carbon atom
C	30	aromatic and conjugated carbon atom
H	90	attached to carbon atom of type 30
H	91	attached to carbon atom of type 1
H	92	attached to carbon atom of type 2
H	94	attached to carbon atom of type 4

One important aspect of KESSHOU is to use a spherical crystal of a finite size, which has to be determined each time, at least until we gain enough experience, for each crystal being considered. In the present work, average intermolecular energy E, root-mean-square average of energy gradients with respect to each of the cell constant $G(A), G(B), \ldots, G(\gamma)$ [*vide infra*] and also the same average with respect to the sum of all cell constants AvG have been computed for the ALL208 dataset while changing the maximum atom-pair distance between the central and the farthest surrounding molecule from 6 to 22 Å at an interval of 1 Å (Fig. 1). Both intermolecular energy and its gradient stay almost constant when the distance exceeded 15 Å. Hence in the subsequent calculations we fixed the radius of crystal sphere at this value.

Gradients of Intermolecular Energy (MM3) for Lattice Constant

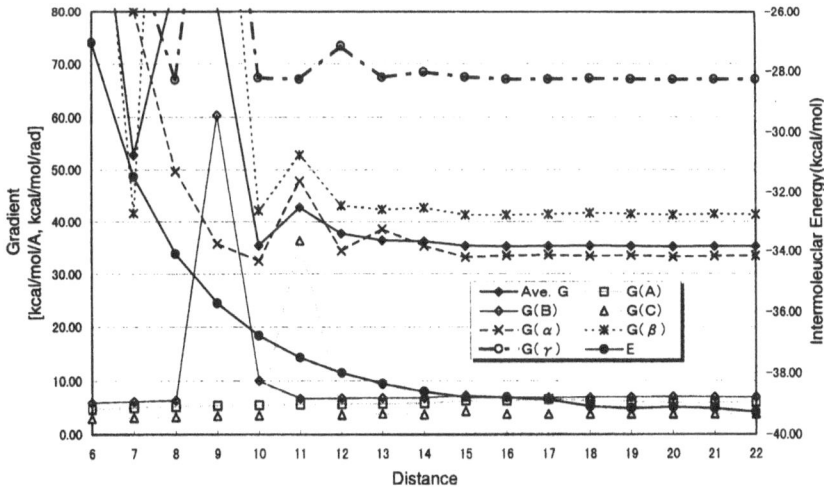

Figure 1. Dependence of the sum of intermolecular energy as well as of the RMS averages of its gradient with respect to cell constants upon the maximum atom-atom distance (Å).

123

Genetic Algorithm

Our new optimization routine KGA97 uses the genetic algorithm (GA),[13] which applies such principles of heredity as mutation, cross-over, and natural selection to chromosomes, represented by bit arrays, and evaluates the best fit to the problem being addressed. As the generation progresses, the choromosomes are improved and the arrays are transformed accordingly. Due to the efficient learning mechanism built in the algorithm, GA is suitable for handling large amounts of data and has been applied successfully to a number of chemical problems.[14] We ported a Fortran GA driver from the public domain of Internet,[15] and added a uniform cross-over routine for real numbers in which the chromosome in question will selectively undergo cross-over with others to achieve higher fitness.

The intermolecular van der Waals constants were then adjusted so that the gradient of intermolecular potential energy E_{inter} with respect to small changes in the cell constants $\mathbf{K}^r$ of the reference crystal structure r takes the smallest possible value.

$$\mathbf{K}^r = \{K_1^r, K_2^r, \ldots, K_6^r\} = \{A^r, B^r, C^r, \alpha^r, \beta^r, \gamma^r\} \tag{2}$$

If we regard a set of van der Waals constants $\mathbf{p}$ as a chromosome, E_{inter} can be defined as a function of both $\mathbf{p}$ and $\mathbf{K}^r$. Hence the convergence criteria can be written as:

$$\frac{\delta E^r(\mathbf{p}, \mathbf{K})}{\delta K_l^r} \cong 0, \quad l = 1, 2, 3, 4, 5, 6 \tag{3}$$

Thus, the fitness of transformed chromosome can be determined by an evaluation function $F(\mathbf{p})$ as follows:

$$F(\mathbf{p}, \mathbf{K}) = \frac{1}{6} \sum_{r=1}^{N_r} \sum_{l=1}^{6} \left(\frac{\delta E^r(\mathbf{p}, \mathbf{K})}{\delta K_l^r} \right)^2 \tag{4}$$

where N_r is the number of crystal data. A smaller $F(\mathbf{p})$ value means a better fit.

We start from the original sets of MM3 van der Waals constants, and they are allowed to change at intervals of $1/2^{12}$ up to a maximum of $\pm 10\%$ of the original.

Determination of GA Parameters

The efficiency of GA optimization depends on the condition-setting parameters in GA. We used here only one crystal data, BAPBID10 (3,4,5:8,9,10-dinaphtho-tricyclo-[5.3.2.2^{2,6}] tetradeca-11,13-diene 1) taken from ALL208 to execute KESSHOU-KGA97 under various sets of GA parameters in order to compare the performance. Pertinent results are shown in Table 2. Starting from the initial fitness of 7.6 for the original MM3 van der Waals constants, GA worked well in all the cases studied to reach very good fitness (low fitness values). Cases 2 and 3 turn out to be the most efficient. Case 3 is the same as Case 2 except that the number of generation was doubled.

Table 2. Fitting Intermolecular van der Waals Constants to the Crystal Structure of **1** under various GA parameters

variable	Case 1	Case 2	Case 3	Case 4	Case 5	Case 6
No. of chromosome	10	20	20	20	20	30
No. of generations	100*2	100	100*2	100	100	100
No. of elite replicns	1	1	1	1	1	1
P.[a] of cross-over	0.500	0.500	0.500	0.500	0.500	0.500
P. of jump mutation	0.025	0.025	0.025	0.025	0.050	0.025
P. of creep mutation	0.250	0.250	0.250	0.250	0.500	0.250
Rate of selection	0.600	0.600	0.600	1.000	1.000	0.600
results						
Final fitness	0.435	0.239	0.190	0.298	0.902	0.280
Generations reqd[b]	187	62	186	46	62	90
Number of evalns	1,870	1,240	3,720	920	1,240	2,700

[a]Probability.
[b]Minimum number of generations required to reach the best fitness.

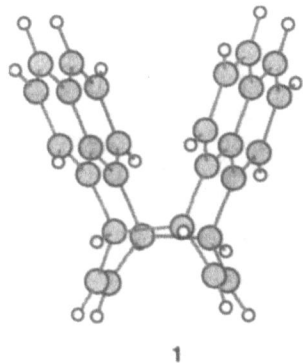

1

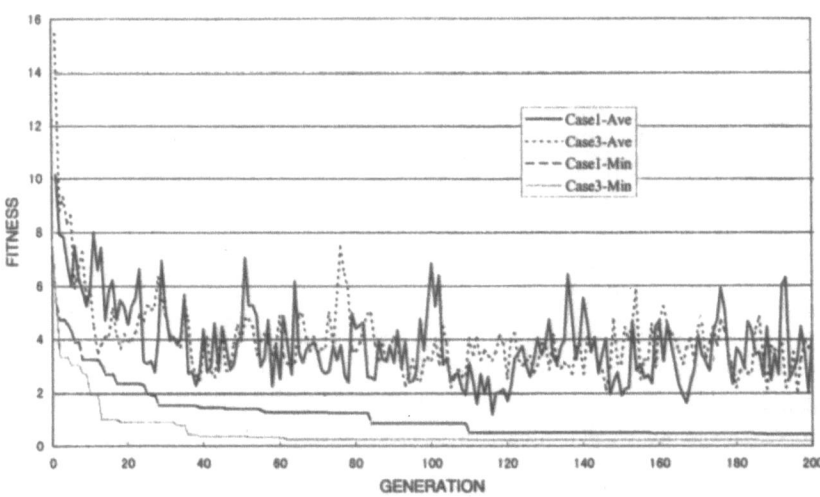

Figure 2. Trajectories of averages and minima of fitness of van der Waals parameters to the crystal structure of 1 with the progress of GA-optimization under conditions defined by Cases 1 and 3 (Table 2).

Figure 2 compares changes in the fitness with the generations between Cases 1 and 3. The latter reached a fitness of 0.24 before the 100th generation was passed, and still improved to 0.19 near the 200th generation. This set proved the best even when other sets containing different number of chromosomes, rate of selection, and mutation probabilities were compared (not included in Table 2).

4. RESULTS

Following the tests mentioned above, we performed GA-optimization of intermolecular van der Waals constants with the parameters set of Case 3, first using ALL208 dataset as the reference. However, not only the initial fitness was very poor (509) but also the convergence was extremely slow: fitness decreased only to 332 at the 46th generation and 330 at the 70th generation. On the other hand, the second run using ALL039 dataset was begun with an initial fitness of 58 and continued to the 145th generation to give a fitness of 34, which stayed unchanged to the 170th generation. Although the improvement in fitness did not change appreciably between the two runs (64.8% in the first vs 58.6 % in the second), we mention below only the results of the second run.

Before starting the optimization runs, we counted the number of intermolecular interaction pairs within the crystal of 15 Å radius at every 1 Å interval for each pairs. The total number is 2.4 million for ALL039 (Table 3) and 14.8 million for ALL208. Examination of Table 3 reveals rather uneven distribution of atom pairs. Especially small are the number of those interactions involving type 4 (acetylenic) carbon atoms. Any results on these pair interactions will depend on the choice of standard data, hence they will not be discussed below.

Similarly, distribution of intermolecular interaction energies at the outset was calculated at every 1 Å interval (Table 4). An interesting fact emerges from this Table: interactions occurring within the interatomic distance range of 3–6 Å account for 78.6 % of the total, within 3–7 Å 88.2 %, and within 3–8 Å 93.9 %. Hence we may say that molecules in the crystal are held together by interactions acting mainly between 3 to 8 Å. Conversely, any observations made outside this range are based on small number of data, hence may be unreliable.

Table 5 gives the original van der Waals constants of MM3 as well as the final values for intermolecular van der Waals interactions fit to ALL039. Changes are in most cases large, many of them almost reaching to the limit of 10 %. New atom pairs generally ended up with values considerably different from the original. Figures 3 to 5 reproduce $C \cdots C$, $C \cdots H$ and $H \cdots H$ interactions curves over a wide range of interatomic distance. It is readily seen that even small changes in the van der Waals constants (Table 5) lead to appreciable shifts of both the depth of well and the optimum distance. In order to check if the changes are reasonable, we studied individual cases more carefully below.

$C \cdots C$ Interactions (Fig. 3). According to the original MM3 scheme, intramolecular van der Waals interaction energies increase in the order $C_{sp^3}(1) \cdots C_{sp^3}(1) < C_{sp^3}(1) \cdots C_{sp^2}(2) < C_{sp^2}(2) \cdots C_{sp^2}(2)$ for the same distance. Namely the depth of van der Waals potential curve increases in this order, or C_{sp^2} atom is softer than C_{sp^3}. This trend is reasonable in view of the higher polarizability of unsaturated carbon atom relative to that of the saturated. However, giving identical van der Waals radius of 2.0 Å to the two types of carbon atoms in the original MM3 is illogical.

In contrast, the trend in the intermolecular interactions obtained in this work is consistent. $C_{sp^3}(1)$ now came out as the very large and hard atom with the well-depth being decreased to one third of the original MM3 value. Van der Waals radius of $C_{sp^3}(1)$ may be readily es-

Table 3. Number of Intermolecular van der Waals Interaction Pairs within Concentric Shells of Unit Thickness(ALL039)

Atom Type Pair	Distance (Å) 2-3	3-4	4-5	5-6	6-7	7-8	8-9	9-10	10-11	11-12	12-13	13-14	14-15	TOTAL
1- 1	0	197	1,291	2,545	4,071	5,033	6,089	8,045	9,852	12,008	14,216	16,459	19,158	98,964
1- 2	0	54	362	814	1,340	1,706	2,252	2,652	3,492	4,136	4,926	5,474	6,588	33,796
1- 4	0	22	56	122	174	212	282	342	402	532	638	744	836	4,362
1- 30	0	480	1,766	3,186	5,090	7,308	9,856	12,042	14,548	17,254	20,712	23,520	27,514	143,276
1- 90	22	550	1,170	1,808	2,824	3,946	5,266	6,396	8,258	9,464	11,326	12,606	15,380	79,016
1- 91	10	2,306	4,280	7,766	11,102	15,616	20,662	26,054	31,310	38,118	45,414	52,744	61,126	316,508
1- 92	0	84	218	358	440	708	916	1,100	1,476	1,694	2,092	2,426	3,010	14,522
1- 94	0	20	46	54	72	124	110	184	180	286	334	396	412	2,218
2- 2	0	117	583	928	1,399	1,952	2,790	3,358	3,892	4,750	5,721	6,897	7,551	39,938
2- 30	0	198	562	836	1,366	1,928	2,528	3,012	3,616	4,380	5,278	6,038	7,118	36,860
2- 90	10	196	390	542	782	1,182	1,406	1,830	2,300	2,788	3,198	3,788	4,306	22,718
2- 91	26	312	740	1,294	2,056	2,948	3,684	4,992	5,780	7,176	8,304	9,554	11,314	58,180
2- 92	22	476	798	1,212	1,872	2,762	3,768	4,422	5,452	6,694	8,082	9,406	10,402	55,368
4- 4	0	37	71	61	107	210	172	343	345	359	460	692	621	3,478
4- 91	0	68	66	164	324	354	414	546	644	956	948	1,088	1,348	6,920
4- 94	16	46	42	56	136	176	212	306	352	396	480	532	690	3,440
30- 30	230	1,248	4,087	6,695	9,999	13,532	18,203	23,103	28,604	33,712	39,829	46,705	54,815	280,532
30- 90	100	3,194	5,446	7,962	12,180	17,646	22,060	28,494	35,336	42,764	51,116	60,096	68,288	354,812
30- 91	14	1,752	3,412	5,346	8,150	12,568	16,650	20,852	25,170	30,416	35,754	40,912	46,990	248,072
30- 92		272	434	746	1,102	1,632	1,964	2,444	2,936	3,712	4,272	5,002	5,886	30,416
90- 90	386	1,210	1,804	2,602	4,040	5,652	7,420	9,514	12,446	14,042	17,394	19,488	23,451	119,449
90- 91	400	1,036	1,802	2,776	4,398	6,506	8,480	10,274	13,104	15,324	18,170	21,192	24,492	127,954
90- 92	52	200	280	378	702	906	1,150	1,448	1,896	2,140	2,788	2,848	3,470	18,258
91- 91	767	1,862	3,550	6,082	9,692	12,990	17,547	22,019	26,774	32,588	38,652	45,064	52,222	269,809
91- 92	54	156	366	560	848	1,232	1,416	1,950	2,554	2,956	3,400	4,140	4,696	24,328
91- 94	14	36	40	98	140	166	186	262	342	426	538	596	674	3,518
92- 92	100	156	300	392	739	993	1,273	1,749	2,096	2,427	2,877	3,504	3,893	20,499
94- 94	3	14	14	11	30	44	60	84	84	93	107	129	192	865
TOTAL	2,226	16,299	33,976	55,394	85,175	120,032	156,816	197,817	243,241	291,591	347,026	402,040	466,443	2,418,076

127

Table 4. Sum of Intermolecular van der Waals Interaction Pairs within Concentric Shells of Unit Thickness(ALL039)

Atom Type Pair	Distance (Å)													TOTAL
	2-3	3-4	4-5	5-6	6-7	7-8	8-9	9-10	10-11	11-12	12-13	13-14	14-15	
1- 1	0.00	-9.74	-41.70	-27.59	-16.32	-8.51	-4.88	-3.30	-2.19	-1.55	-1.12	-0.81	-0.62	-118.32
1- 2	0.00	-2.73	-11.81	-9.54	-5.62	-3.04	-1.94	-1.16	-0.83	-0.57	-0.41	-0.29	-0.23	-38.15
1- 4	0.00	-1.03	-1.61	-1.51	-0.78	-0.40	-0.25	-0.15	-0.10	-0.07	-0.05	-0.04	-0.03	-6.01
1- 30	0.00	-23.70	-57.51	-37.41	-21.66	-13.08	-8.40	-5.26	-3.46	-2.38	-1.73	-1.24	-0.94	-176.78
1- 90	0.14	-24.14	-19.51	-9.62	-5.39	-3.24	-2.03	-1.26	-0.89	-0.59	-0.43	-0.30	-0.24	-67.49
1- 91	-0.14	-104.48	-70.19	-40.37	-21.23	-12.73	-7.93	-5.14	-3.38	-2.38	-1.72	-1.26	-0.95	-271.91
1- 92	0.00	-3.38	-3.76	-1.79	-0.87	-0.59	-0.35	-0.22	-0.16	-0.11	-0.08	-0.06	-0.05	-11.42
1- 94	0.00	-0.93	-0.78	-0.27	-0.13	-0.11	-0.04	-0.04	-0.02	-0.02	-0.01	-0.01	-0.01	-2.35
2- 2	0.00	-5.23	-19.66	-11.28	-6.35	-3.72	-2.49	-1.55	-0.99	-0.71	-0.50	-0.39	-0.28	-53.16
2- 30	0.00	-9.39	-18.28	-10.30	-6.11	-3.70	-2.28	-1.39	-0.92	-0.64	-0.47	-0.34	-0.26	-54.07
2- 90	0.58	-8.29	-7.57	-3.30	-1.74	-1.14	-0.64	-0.43	-0.29	-0.21	-0.14	-0.11	-0.08	-23.34
2- 91	1.02	-13.66	-13.81	-7.98	-4.65	-2.85	-1.69	-1.16	-0.73	-0.53	-0.37	-0.27	-0.21	-46.89
2- 92	0.75	-20.27	-15.41	-7.59	-4.24	-2.67	-1.70	-1.02	-0.70	-0.49	-0.36	-0.27	-0.19	-54.16
4- 4	0.00	-1.62	-2.45	-0.69	-0.49	-0.40	-0.16	-0.16	-0.09	-0.05	-0.04	-0.04	-0.02	-6.21
4- 91	0.00	-3.01	-1.22	-1.00	-0.73	-0.35	-0.19	-0.12	-0.08	-0.07	-0.04	-0.03	-0.02	-6.87
4- 94	1.92	-2.07	-0.98	-0.35	-0.33	-0.17	-0.09	-0.07	-0.05	-0.03	-0.02	-0.02	-0.01	-2.28
30- 30	0.00	-59.52	-135.39	-81.71	-46.07	-25.81	-16.44	-10.65	-7.27	-4.95	-3.54	-2.62	-2.00	-395.96
30- 90	16.99	-136.34	-107.90	-49.98	-27.24	-16.97	-10.04	-6.60	-4.51	-3.16	-2.29	-1.69	-1.25	-350.97
30- 91	6.51	-74.65	-65.32	-33.02	-18.41	-12.09	-7.56	-4.85	-3.20	-2.24	-1.60	-1.15	-0.86	-218.45
30- 92	0.84	-11.85	-8.53	-4.65	-2.41	-1.57	-0.89	-0.57	-0.37	-0.27	-0.19	-0.14	-0.11	-30.70
90- 90	5.46	-42.10	-17.42	-7.26	-4.15	-2.51	-1.54	-1.00	-0.72	-0.47	-0.35	-0.25	-0.20	-72.50
90- 91	4.13	-34.92	-16.57	-7.92	-4.56	-2.86	-1.75	-1.09	-0.76	-0.51	-0.37	-0.27	-0.20	-67.64
90- 92	-0.01	-6.74	-2.54	-1.04	-0.71	-0.40	-0.24	-0.15	-0.11	-0.07	-0.06	-0.04	-0.03	-12.15
91- 91	3.78	-64.53	-32.98	-17.31	-10.03	-5.72	-3.61	-2.34	-1.55	-1.09	-0.78	-0.58	-0.43	-137.19
91- 92	0.48	-5.23	-3.29	-1.51	-0.87	-0.53	-0.29	-0.21	-0.15	-0.10	-0.07	-0.05	-0.04	-11.85
91- 94	-0.51	-1.02	-0.36	-0.28	-0.14	-0.07	-0.04	-0.03	-0.02	-0.01	-0.01	-0.01	-0.01	-2.50
92- 92	-2.04	-5.50	-2.99	-1.14	-0.75	-0.42	-0.26	-0.18	-0.12	-0.08	-0.06	-0.04	-0.03	-13.61
94- 94	-0.16	-0.67	-0.16	-0.02	-0.03	-0.02	-0.01	-0.01	0.00	0.00	0.00	0.00	0.00	-1.10
TOTAL	39.74	-676.73	-679.68	-376.42	-212.00	-125.65	-77.74	-50.10	-33.67	-23.38	-16.83	-12.29	-9.28	-2204.54

Table 5. Initial (MM3, Left) and Final (150th Generation, Right) Parameters of Intermolecular van der Waals Constants (ALL039)

Atom Type		MM3(92) FITNESS = 57.780			ALL039 (150th Generation) FITNESS = 34.144		
I	J	A	B	C	A	B	C
1	1	4968.00	-2.9412	-280.225	5411.91	-2.7443	-259.716
1	2	7154.74	-3.0000	-358.359	6694.70	-2.8377	-371.039
1	4	7154.74	-3.0151	-347.742	6904.72	-3.1102	-347.343
1	30	7154.74	-3.0000	-358.359	7223.75	-2.8455	-334.057
1	90	4232.00	-3.9267	-42.153	3912.97	-3.8363	-39.978
1	91	4232.00	-3.9267	-42.153	4355.50	-4.1834	-38.712
1	92	4232.00	-3.9267	-42.153	4607.45	-3.8411	-42.368
1	94	4232.00	-3.9267	-42.153	4244.92	-3.9483	-44.050
2	2	10304.00	-3.0612	-457.180	10623.31	-3.2374	-418.652
2	30	10304.00	-3.0612	-443.362	11035.97	-3.0366	-422.087
2	90	4232.00	-3.9267	-42.153	4301.34	-4.0265	-43.814
2	91	4232.00	-3.9267	-42.153	3930.54	-3.5831	-40.682
2	92	4232.00	-3.9267	-42.153	3905.94	-3.6351	-38.253
4	4	10304.00	-3.0928	-429.894	10367.16	-3.1125	-405.066
4	91	4232.00	-3.9267	-42.153	4583.27	-4.2590	-40.481
4	94	4232.00	-3.9267	-42.153	4112.22	-3.6905	-42.817
30	30	10304.00	-3.0612	-457.180	9344.56	-3.0780	-417.491
30	90	4232.00	-3.9267	-42.153	3887.34	-3.6073	-38.918
30	91	4232.00	-3.9267	-42.153	3978.70	-4.0415	-44.969
30	92	4232.00	-3.9267	-42.153	4510.31	-3.5939	-39.910
90	90	3680.00	-3.7037	-52.057	3396.11	-3.7572	-47.703
90	91	3680.00	-3.7037	-52.057	3894.15	-3.9321	-48.540
90	92	3680.00	-3.7037	-52.057	3676.68	-3.8793	-54.820
91	91	3680.00	-3.7037	-52.057	3421.46	-3.7906	-50.663
91	92	3680.00	-3.7037	-52.057	3774.63	-3.8328	-55.305
91	94	3680.00	-3.7037	-52.057	3528.22	-3.5178	-53.882
92	92	3680.00	-3.7037	-52.057	3686.74	-3.6495	-52.069
94	94	3680.00	-3.7037	-52.057	3497.12	-4.0142	-47.795

timated by dividing the 1–1 distance at the minimum by 2, namely 2.4 Å, which represents 20 % increase from the original MM3 value. The intermolecular van der Waals radii thus estimated decrease in the order $C_{sp^3}(1)$, 2.4 Å > $C_{arom}(30)$, 2.0 Å > $C_{sp^2}(2)$, 1.7 Å, and the well-depths increase in the reverse order. The optimum van der Waals distance of 3.4 Å obtained here for $C_{sp^2}(2) \cdots C_{sp^2}(2)$ interaction agrees perfectly well with the interlayer distance of graphite (3.3539 Å).[16] Together with the large well-depth, this feature must be responsible for the often recognized tendency of aromatic rings to stack face-to-face in crystals.

The $C \cdots C$ potential curves change wildly in the important distance range of 3–5 Å, therefore these interactions must be of dominant importance in the crystal cohesion. Reversal in the order of 2–30 and 30–30 curves is the only minor inconsistency observed in this part of analysis.

H $\cdots$ H Interactions (Fig. 4). Compared to $C \cdots C$ interactions, differences among pair types are small, but still appear significant. Estimating van der Waals radii of three different types of hydrogen atom, $H(92)[-C_{sp^2}(2)]$, $H(90)[-C_{arom}(30)]$, and $H(91)[-C_{sp^3}(1)]$ as above, we obtain 1.6_5, 1.5_5, and 1.5_0 Å, respectively. This order can be interpreted to mean that hydrogen atom becomes harder as it is bound to more electropositive atome like C_{sp^2}. Up to this point, the analysis has gone smooth, but breaks down below.

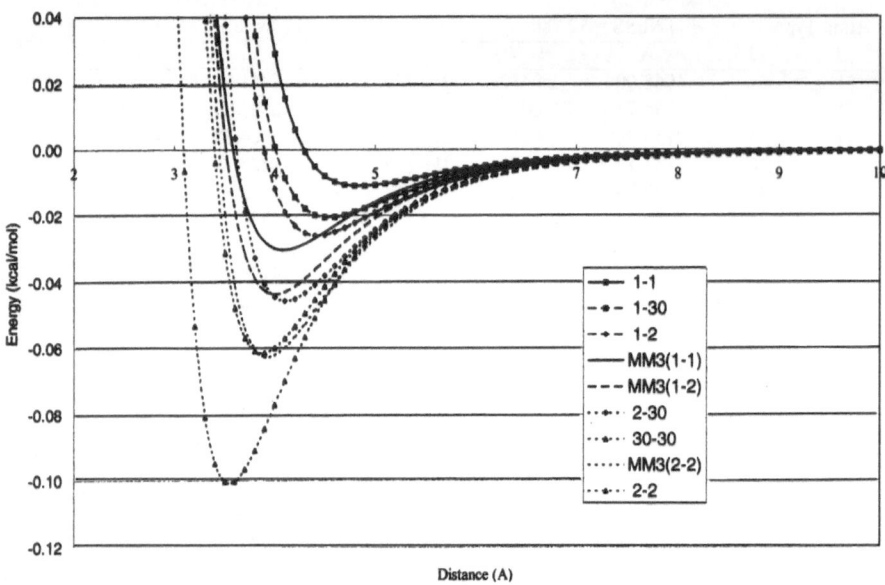

Figure 3. Van der Waals potential curve of intermolecular C⋯C interactions.

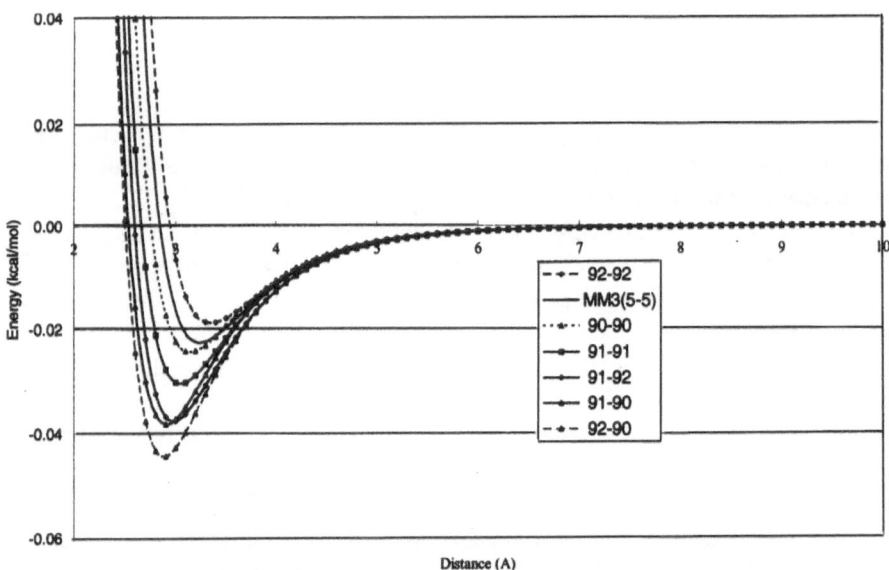

Figure 4. Van der Waals potential curve of intermolecular H⋯H interactions.

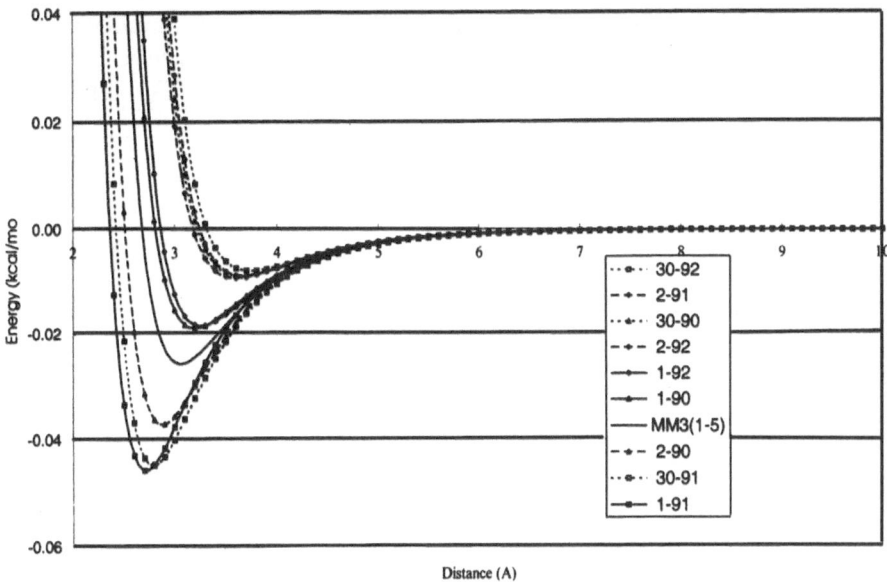

Figure 5. Van der Waals potential curve of intermolecular C···H interactions.

C···H Interactions (Fig. 5). For the heterogeneous interaction C···H, we tested the sum rule of van der Waals radii using the values obtained from the homogeneous pairs, C···C and H···H. Table 6 summarizes the results, which reveal no serious discrepancy, between the calculated and observed (Figs. 3–4), and give consistent orders (left block columns).

However, the sum rule failed for the heterogeneous C···H interactions (right block columns). The first indication of failure comes from the pairs involving $C_{sp^3}(1)$, the most common pair in the crystals of saturated hydrocarbons: the first three lines all show much shorter optimal distances than the sum rule indicates. This observation contradicts to all our previous knowledge. Still more significant is the failure to detect the so-called "C−H···π interaction", in which unsaturated carbon atom interacts with H to form hydrogen bond-like complex C−H···$C_{arom}(30)$, $C_{sp^2}(2)$.[17] These pairs (italicized in the right block columns of Table 6) are expected to shift the optimum distances to shorter values than the calculated, but only the 30–91 and 2–90 did so, while all other pairs studied indicate considerable increase in the distance!

5. DISCUSSION

Missing Electrostatic Interaction

We interpret the above results to have arisen from the neglect of electrostatic interactions in the hydrocarbon force field of MM3. Unlike the AMBER/CHARMm line of force fields, MM series of Allinger and several similar schemes lack electrostatic term for neutral hydrocarbons, hence it is likely that the contributions from electrostatic interactions are somehow buried in other terms. This neglect was perhaps not truly critical as long as the MM calculations are used for an isolated single molecule in vacuum, because most of the hydrogen atoms are located on the outer surface of molecule and only a few of them enter into the important

Table 6. Estimated and Observed van der Waals Distance of Largest Attraction in C⋯C, H⋯H (homogeneous) and C⋯H (heterogeneous) Interactions

Homogenous interaction			Heterogeneous interaction		
Pair	Optimum distance, Å		Pair	Optimum distance, Å	
	Calc	Obs		Calc	Obs
1-1		2.4	1-92	4.0_5	3.3
30-30		2.0	1-90	3.9_5	3.2
2-2		1.7	1-91	3.9_0	2.7
1-30	4.4	4.5	30-92	3.6_5	3.9
1-2	4.1	4.3	30-90	3.5_5	3.8
2-30	3.7	4.0	30-91	3.5_0	2.8
92-92		1.6_5	2-92	3.3_5	3.6
90-90		1.5_5	2-90	3.2_5	2.9
91-91		1.5_0	2-91	3.2_0	3.7
92-91	3.1_5	3.0			
92-90	3.2_0	2.9			
90-91	3.0_5	2.9			

distance region of nonbonded interaction (3–6 Å) with carbon atoms in the other parts of molecule. Electrostatic energy must be large in 1,3– and 1,4–H/C nonbonded interactions, hence difficult to overlook. Here again they can be readily absorbed into the angle bending constants of the type H−C−C and the torsional constants of the type H−C−C−C, respectively. This kind of latent defect, if any at all, is hard to detect unless the MM scheme is applied to the intermolecular interactions.

On the other hand, small charge separation must exist for any heteropolar bonds involving C−H, C_{sp^3}−C_{sp^2} and C_{sp^3}−C_{sp}, as has been amply borne out by quantum chemical calculations.[18] Working over long distances and producing large energies, the electrostatic interactions should play a dominant role in the cohesion of crystals even in hydrocarbons.

This interpretation does not contradict with the consistent parameters set obtained for homogeneous C⋯C and H⋯H interactions. Electrostatic interactions in these pairs always work between the same sign, hence affect the van der Waals potential curve in the form of parallel upward shift. Such an effect must have been absorbed into van der Waals constants in the course of GA optimization. However, when it comes to the heterogeneous pairs C⋯H, electrostatic interactions are now not repulsive but attractive. Therefore the same constants as used for the homogeneous interactions cannot be used anymore. This is the most likely reason why the sum rule of van der Waals radii failed for C⋯H interactions.

The hypothesis of missing electrostatic interactions also explains the large change in the MM3 van der Waals constants of hydrocarbons that we observed when they were re-optimized on the dataset of hydrocarbon crystals. Otherwise, a parameters set optimized for the isolated molecules should work for crystals as well, as long as the atom-atom pair potential paradigm remains valid. Because of possible contamination of the electrostatic effects in the van der Waals constants, the differential van der Waals radius of carbon atom due to hybridization as well as that of hydrogen attoached to differently hybridized carbon atom mentioned above will not be discussed here.

Perspectives

Finally, a perspective on studying the very weak intermolecular interactions in crystal will be briefly discussed. Directional, hydrogen bond-like interactions like CH⋯O, and CH⋯Cl have been often studied based on statistical analysis using CSD.[19] The sum rule criterion presented above for CH⋯π interaction may also be applied to these additional types. Such an analysis will provide not only geometric informations but also energetic evaluation of the interactions. Indirect as it is, this approach seems to be the only available method to date in view of the weakness of molecular orbital method for such a purpose. However, we should be aware of the possible artefact that may occur if the new interactions are always absorbed into the van der Waals term while fixing the atomic chages.

Work is in progress to include multipole electrostatic charges and resume the GA-optimization of van der Waals constants, and perhaps to construct a new MM scheme based entirely on CSD.

6. SUMMARY

1. An attempt was made to optimize intermolecular van der Waals constants by using the MM3 hydrocarbon force field to fit large sets of crystal structures of hydrocarbons taken from the Cambridge Structural Database.

2. Preliminary optimization runs by genetic algorithm were stopped after the fitness values improved by about 60%. Then it was found that the original MM3 van der Waals constants changed to a large extent. New constants for homogeneous interaction pairs C···C and H···H showed acceptable consistency, but those for heterogeneous pairs C···H did not.

3. The lack of electrostatic interaction terms in the MM3 force field is considered responsible for the failure in optimizing the MM3 van der Waals constants for intermolecular interactions.

Acknowledgements

We thank participants of INDABA2 for many useful comments, especially Dr. F. J. J. Leusen for supplying us with his latest results.

REFERENCES

1. F. H. Allen, *in*: "Fundamental Principles of Molecular Modeling," W. Gans, A. Amann, and J. C. A. Boeyens, eds., p. 105, Plenum Press, New York (1996).
2. (a) J. R. Hill, and J. Sauer, *J. Phys. Chem.* 99:9536 (1995). (b) B. J. Palmer and J. L. Anchell, *J. Phys. Chem.* 99:12239 (1995). (c) D. Q. Mcdonald and W. C. Still, *J. Org. Chem.* 61:1385 (1996). (d) E. A. Mash, T. M. Gregg, and M. T. Stahl, *J. Org. Chem.* 62:3715 (1997). (e) G. Y. Liang, X. N. Chen, J. A. Dustman, A. H. Lewin, and J. P. Bowen, *J. Comput. Chem.* 18:1371 (1997).
3. Y. Fukazawa, and T. Haino, *J. Syn. Org. Chem. Jpn.*, 54:419 (1996).
4. A. Gavezzotti, ed. "Theoretical Aspects and Computer Modeling of the Molecular Solid State," John Wiley & Sons, Chichester (1997).
5. E. Ōsawa, Potential Energy Calculations of Crystals, *in*: "Reactivity in Molecular Crystals," Y. Ohashi, ed., Kodansha, Tokyo/VCH Publ., New York (1993).
6. The algorithm was used by Sironi to study crystal dynamics of iron carbonyl: A. Sironi, *Inorg. Chem.*, 35:1725 (1996).
7. BIGSTRN3: R. B. Nachbar, Jr., and K. Mislow, *QCPE*, #514.
8. BS/KESSHOU (version 1.0): E.Ōsawa, P. M. Ivanov, H. Gotō, M. Yamato, J. Rudzinski, and A. Aoki, *JCPE*, No. P111. Japan Chemistry Program Exchange: 1-7-12 Nishinenishi, Tsuchiura, Ibaraki-ken 300, Japan, Fax 81-298-30-4162, homepage http://jcpe.ocha.ac.jp/.
9. U. Burkert and N. L. Allinger. "Molecular Mechanics," American Chemical Society, Washington, D. C. (1982).
10. N. L. Allinger, Y. H. Yuh, and J. H. Lii, *J. Am. Chem. Soc.*, 111:8551 (1989).
11. P. M. Ivanov, *J. Mol. Struct.* in press.
12. (a) Y.-M. Xun, T. Ouchi, C. Jaime, E. Ōsawa, A. Okamoto, and T. Higuchi, *JCPE Newsletter*, 1:24 (1989). (b) T. Murakami, E.Ōsawa, and Y. Mochizuki, *QCPE Bull.*, 8:15 (1988). (c) E.Ōsawa, and Y. Mochizuki, *QCPE Bull.* 5:119 (1985).

13. D. E. Goldberg. "Genetic Algorithm in Search, Optimization and Machine Learning," Addison-Wesley, Reading, MA (1989).

14. (a) D. L. Carroll, *in*: "Developments in Theoretical and Applied Mechanics," Vol. XVIII, H. Wilson, R. Batra, C. Bert, A. Dawis, R. Schapery, D. Stewart, and F. Swinson., eds., University of Alabama, Birmingham (1996), p. 411. (b) J. Devillers, ed. "Genetic Algorithms in Molecular Modelling," Academic Press, New York (1996).

15. D. L. Carroll, FORTRAN Genetic Algorithm (GA) Driver, http://www.staff.uiuc.edu/~carroll/ga.html. See also ref. 14a.

16. M. S. Dresselhaus, G. Dresselhaus, and P. C. Eklund. "Science of Fullerenes and Carbon Nanotubes," Academic Press, San Diego (1996), p. 18.

17. M. Nishio, Y. Umezawa, M. Hirota, and Y. Takeuchi, *Tetrahedron*, 51:8665 (1995).

18. W. J. Hehre, L.Radom, P. v. R. Schleyer, and J. A. Pople. "Ab Initio Molecular Orbital Theory," John Wiley & Sons, New York (1986).

19. (a) G. R. Desiraju, and B. N. Murty, *Chem. Phys. Lett.*, 139:360 (1987). (b) G. R. Desiraju, *J. Chem. Soc., Chem. Commun.*, p. 179 (1989). (c) T. Steiner, and W. Saenger, *J. Am. Chem. Soc.*, 115:4540 (1993). (d) T. Steiner, *J. Chem. Soc., Chem. Commun.*, p. 101 (1994).

REACTIVITY IN SOLID STATE AND ELECTRON DEFORMATION DENSITY DETERMINATIONS

Carl Krüger, K. Angermund, B. Bartkowska, J. Bruckmann, K. H. Claus, J. Kuhnigk, F. Lutz and I. Ortmann

Max-Planck-Institut für Kohlenforschung
Strukturchemie
Mülheim an der Ruhr
Germany

ABSTRACT

This contribution consists of three related topics. After some introductory remarks about the techniques used for Electron Deformation Density Determinations (EDD) the next part is concerned with the solid state photochemistry of styryl-α-pyrone and their isoelectronic nitrogen substituted isomorphs. Investigations of the EDD provide an explanation for the change in photoreactivity, since different forms of conjugation within the molecules as well as different π-π-interactions in the solid state are present.

In the second part of the contribution results on the structural and reactivity investigations of several organic di- and polyenes which are liquids at ambient temperature will be presented. A short description of the techniques of *in situ* crystal growing used at low temperature will be followed by a summary of recent results. The photoreactivity of single crystals as grown on a diffractometer has been directly investigated and analysed using various techniques.

Similar crystal growing techniques have been used to characterize numerous substituted phosphines in the solid state. The derived results as presented in the third part of the paper will give a more detailed insight into the chemical properties of these ligands which are popular in transition metal catalysis.

1. INTRODUCTION

Technical achievements of the last decades of this century have introduced a new era of structural chemistry. During recent years structural chemistry of the solid state was divided into chemical crystallography, inorganic, small-molecule (organic) and biological structural chemistry. Practical crystallography itself has changed tremendously with the advent of modern radiation sources and diffraction equipment, such as area counters (CCD techniques) and imaging plates. This fast equipment enables laboratories, in conjunction with present-day software and fast, reasonably priced computing equipment, to do routine structure determinations within a short period of time, and even the scientists involved in this method are going to be replaced by less trained personnel. These are but a few reasons why structural chemistry has to cover, in addition, new topics in the future. Amongst these are extensive use of powder diffraction methods using various radiation sources, new crystal growing techniques, the use of data bases to investigate the vast amount of information by statistical means, the use of computational chemistry (from molecular modeling to *ab initio* methods) in structural chemistry, and finally to improve structural investigations by using high resolution techniques. This contribution is concerned with only a few of these topics from this list of achievements.

2. THE USE OF HIGH RESOLUTION DATA IN SOLID STATE CHEMISTRY

Topochemical reactions gained interest already at the turn of the century.[1-6] However, it was not until 1943 that it was recognized that reactions in the solid state are controlled by the arrangement of molecules in the crystal lattice.[7] Since many years we have been interested in structural investigations of photoreactive solid state systems. These systems are for the most part highly selective and, in many cases, enable access to unusual compounds which differ from the results of solution photochemistry. Solid state photochemistry in general has been a topic for numerous and comprehensive review articles during the recent years.[8,9-16] However, for the interpretation of such solid state reactions the relationship between packing of the molecules in the crystal and the resulting photoproducts is very important. This relationship was first studied by G.M.J. Schmidt *et al.*[8] in their pioneering work on the solid state chemistry of cinnamic acids, and these investigations led finally to the so called *topochemical postulate* (for [2 + 2] cycloaddition reactions in solid state). The important message of this postulate was that reactions between two reactive centers in crystals take place with minimal atomic and molecular movement, and that these centers have to be arranged in distances of about 3.5–4.7 Å. However, reactions have also been observed later which take place with considerable reorganization of the crystal lattice. These results finally led to an extension of the topochemical postulate by M. Cohen *et al.*:[8]

> "Lattice controlled reactions proceed with minimal distortion of the surface of a reaction cavity which contains molecules surrounding the transition state of one photochemically excited molecule or its complex with its neighbor (excimer)."

Recently we have applied in our laboratory not only molecular modeling techniques (CAMD) to explain the results of certain topologically controlled reactions,[17] but also high resolution X-ray techniques (Electron Deformation Density Determinations EDD) to gain information about certain solid state reactions.[18]

Since the standard diffraction techniques used for X-ray structure determination depend on the interaction of X-rays with the electrons of atoms in molecules in a crystal lattice,[19] the diffraction patterns obtained experimentally are directly related to the distribution of electrons in the crystal and thus to the electron density distribution in the molecules. It was our intention to use this information in solid state chemistry.

The method of experimental deformation density determinations was essentially established already in the late sixties of this century,[20] and it has further been developed to the present days. Several reports[21] have been published since then. As the determination of the spatial arrangement of the electron density in molecules might provide a direct route to a better understanding of the chemical bond, numerous theoretical[22] as well as experimental studies[23] have been carried out during the last decade to determine the changes in electron density, which result from redistribution of the valence electrons during formation of chemical bonds. Experience shows that the quality of data as obtained by X-ray and neutron scattering experiments determines essentially the results of a high resolution structure determination.

3. SOME EXPERIMENTAL ASPECTS OF HIGH-RESOLUTION X-RAY STRUCTURE DETERMINATION

The total electron density obtained in a carefully performed X-ray structure determination is biased by the dominating core electrons, as is shown for a *t*-butyl-substituted anthracene **1** (Fig. 1).[24] The total electron density, nevertheless contains complete information about the nature of the bonding in the molecule.

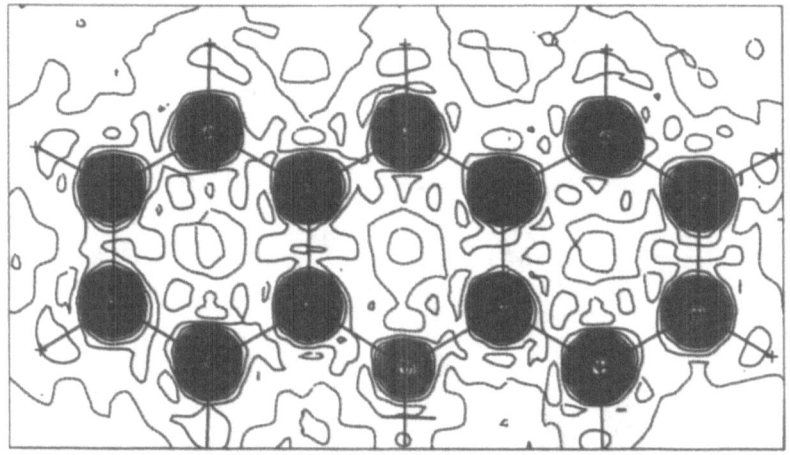

Figure 1. Total electron density of **1**.

In order to present this information in a more useful and visible representation the concept of electron deformation density (EDD) has been introduced . The EDD represents the change in the electron density that results from the deformation of the spherical charge distribution around free atoms when chemical bonds are formed. Since it is the electrons in the valence shells that are mainly redistributed during the deformation, they contribute most to the EDD. The spherically symmetrical electron density distribution for a free atom can be accurately calculated by theoretical methods, and is subsequently used to calculate the scattering factors used in the standard X-ray structure determination. The sum of all the electron densities of the spherical atoms placed at bonding distances from one another is known as the promolecule density, and the molecule so formed is called the "promolecule".[24c]

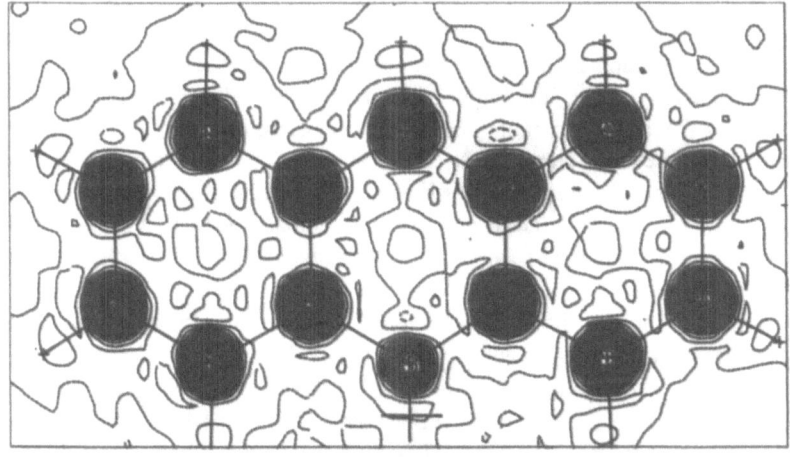

Figure 2. The promolecule density distribution in **1**.[25]

137

The difference between the experimentally determined or theoretically calculated total electron density, which describes the equilibrium state, and the promolecule density, which corresponds to a hypothetical distribution of electrons, is the EDD. A positive deformation density at a position in a molecule means that there is an increase in electron density over and above the sum of the electron densities of the free, spherically symmetrical atoms. Conversely, a negative deformation density represents a reduction in the corresponding electron density.

Figure 3 shows the deformation density in 9-*tert*-butylanthracene.[24c] A positive EDD is found between all bonded atoms. The results of numerous experimental and theoretical studies on a variety of compounds have shown that, depending on the type of atom and bond, not only positive but also negative deformation densities can be observed between bonded atoms.

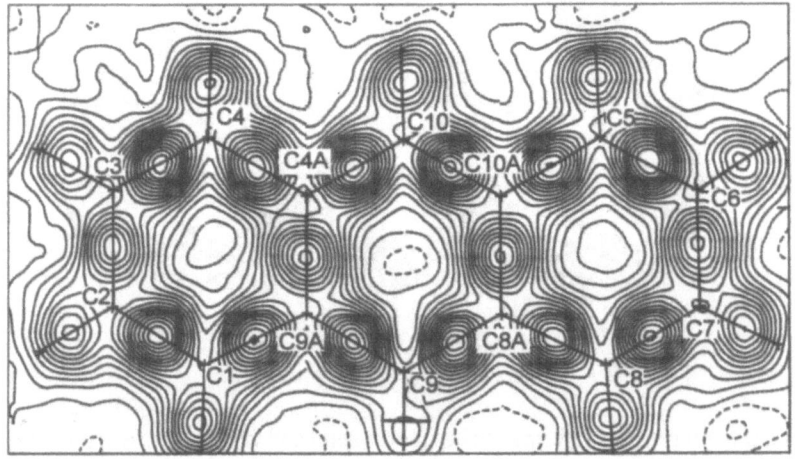

Figure 3. Deformation density in **1**. Contour lines at 0.1 $e/\text{Å}^3$.

The fact that positive EDD is not observed between some types of atoms, particularly electronegative elements,[26] does not necessarily imply that there is no bond between these atoms, but rather reflects the difference in the valence configuration between the bonded and the free atom. This can be qualitatively explained by a simple model. When elements such as carbon, nitrogen, oxygen, or fluorine come together and form bonds, their valence configurations change. In the ensuing process electron density can be thought of as being transferred from atomic orbitals to hybrid orbitals situated further away from the atomic nucleus. Neglecting this fact has caused considerable confusion, as it was thought frequently that other, hitherto unknown causes may be responsible for this observation.

Table 1 gives the approximate electronic configurations of several elements in their free and oriented (bonded) states. The configurations of the oriented atoms will depend on the type of bonding, and those given in Table 1 serve only as examples.

For the electronegative elements (N, O, F), the transition from spherically symmetric free atoms to bonded atoms also results in the formation of lone pairs. These processes are reflected in the EDD. The EDD in the bonding and lone pair regions can be derived by subtracting the population of a *p*-orbital in the free atom from that of the corresponding orbital in the oriented atom[27] (see Table 1).

According to Table 1, the C−C bond has a positive deformation density, whereas the EDD decreases continuously for N−N, O−O, and F−F. A comparison of the results from such a model with the experimentally determined or theoretically calculated EDD shows that in general it is possible not only to obtain the correct sign of the EDD but also a rough estimate of its value. Conversely, starting from the EDD it is possible to obtain information about the electronic configuration of the constituent atoms.

Table 1. Simple method for estimating the EDD in the bonding and lone pair regions of molecules containing main group elements. The electron configuration of the oriented atoms were taken from the following molecules or groups: C, ethane; N, hydrazine; O, hydrogen peroxide; F, molecular fluorine; P, PN; S, CS

| Element | Electronic configuration | | Differences (B − A) in | |
	Free atom (A)	Oriented atom (B) in molecule	Bond region	Lone pair region
C	$1s^2 2s^2 p_x^{2/3} p_y^{2/3} p_z^{2/3}$	$1s^2 2(sp^3)^1 (sp^3)^1 (sp^3)^1 (sp^3)^1$	+1/3	−
N	$1s^2 2s^2 p_x^1 p_y^1 p_z^1$	$1s^2 2(sp^3)^1 (sp^3)^1 (sp^3)^1 (sp^3)^2$	0	+1
O	$1s^2 2s^2 p_x^{4/3} p_y^{4/3} p_z^{4/3}$	$1s^2 2(sp^3)^1 (sp^3)^1 (sp^3)^2 (sp^3)^2$	−1/3	+2/3
F	$1s^2 2s^2 p_x^{5/3} p_y^{5/3} p_z^{5/3}$	$1s^2 2s^2 p_x^1 p_y^2 p_z^2$	−2/3	+1/3
P	$1s^2 2s^2 p^6 3s^2 p_x^1 p_y^1 p_z^1$	$1s^2 2s^2 p^6 3s^2 p_x^1 p_y^1 p_z^1$	0	0
S	$1s^2 2s^2 p^6 3s^2 p_x^{4/3} p_y^{4/3} p_z^{4/3}$	$1s^2 2s^2 p^6 3s^2 p_x^1 p_y^1 p_z^2$	−1/3	+2/3

In heteronuclear bonds (e.g., C−F) the more electronegative element contributes most to the total electron density and thus appears to determine the EDD. This explains, for example, why bonds involving nitrogen atoms have only slightly positive deformation densities while those involving fluorine have negative values.

Calculated deformation densities for species containing elements from the third row of the periodic table (e.g., P or S) show that their bonds have smaller electron deformation densities than the corresponding bonds between lighter elements of the same group. These EDDs can be understood in terms of the small amount of hybridization of the heavier elements in the bonded state. Further predictions or interpretation of the spatial arrangement of the EDD, particularly in polar bonds and in the region close to transition metals, require consideration of additional effects, such as electronegativity and orbital symmetry. Accordingly, comparison of the magnitude of the EDD in bonds between different elements is not possible and the conclusions drawn from such comparisons may be wrong. It is just as meaningless to compare the EDD of a C−F bond with that of a C−C bond as it is to compare their bond lengths. However, if these aspects are taken into consideration[27] then a detailed analysis of the experimentally determined EDD provides a good description of the nature of the bonding in a molecule in a low-energy state in the crystal lattice; on the other hand, an interpretation of the theoretically calculated EDD affords a description of the nature of the bonding in the free molecule.

During the course of every X-ray structure determination, a structural model, consisting of spherically symmetrical non-interacting atoms, is optimized such as to give the best possible least squares fit between the observed diffracted intensities and the intensities calculated for the model. The total electron density is then calculated by assigning phases, determined from the model, to the measured intensities of the reflections. The difference between the observed electron density and the electron density for the model obtained in this way is known as the residual electron density. In favorable circumstances, its spatial distribution allows one already to draw conclusions about the electron density distribution in the molecules.

The atomic positions obtained during the refinement process are, however, influenced by the bonding electrons. This effect is particularly pronounced for light atoms, and is clearly illustrated by a comparison of the average C−H bond distances obtained from X-ray diffraction (0.95 Å) with those determined by neutron diffraction (1.08 Å). A more exact determination of the positions of atoms is possible using one of two experimental techniques. In the first method the atomic positions are determined using only the X-ray diffraction data by taking advantage of the difference in the scattering ability of the core and valence electrons. A disadvantage of this method is that the determination of the positions of the lighter atoms is more inaccurate than that of the exact positions of atoms which are determined by neutron diffraction. This technique is referred to as the X-X method, and it is historically the first entry into this new field.[28] Later it was shown that accurate atomic positions can better be obtained from neutron scattering experiments. This technique is known as the X-N method.[29] It requires considerable experimental effort and much larger crystals than are needed for the X-ray analysis. Since an additional X-ray structure determination is necessary anyway in order to obtain information about the electron density distribution, the X-X method became more popular during the last decades. Problems arise also from the difficulty of assimilating the two sets of results, which are necessarily obtained under slightly different experimental conditions.

Even at present time the EDD determination of compounds that crystallize in acentric space groups is only possible with extreme difficulty, regardless of which of both methods is used.[27] New refinement procedures are of particular interest not only to derive particular physical descriptors, but also for refinement of parameters of molecules in acentric or polar space groups, as all the information necessary to derive meaningful EDD is biased by erroneous phases induced by the model. We have studied one example in order to gain information about this artifact. For 9-methoxyanthracene **2** in space group $Pca2_1$, it was possible with only one anomalous scatterer to assign the correct polar axis. In this case the multipole expansion gave not only correct static model deformation densities, but even meaningful dynamic densities[30] as shown in Fig. 4 and Fig. 5.

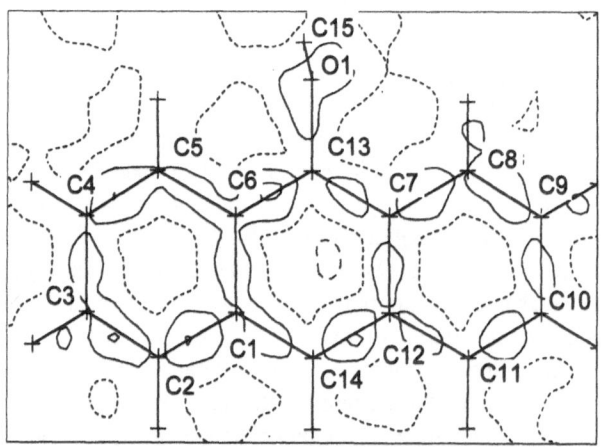

Figure 4. Dynamic deformation density of **2**.

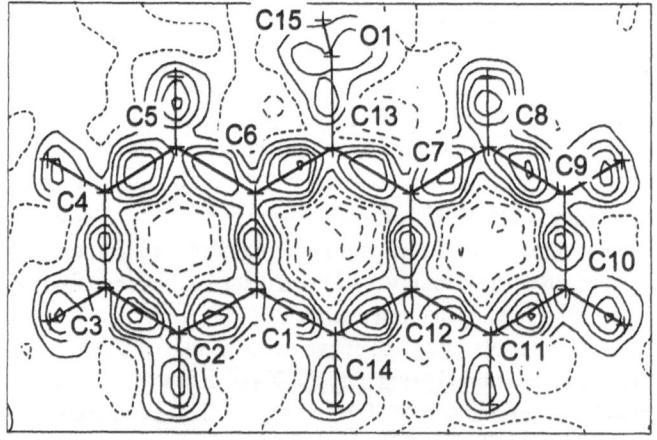

Figure 5. Dynamic deformation density of **2** after multipolar refinement.

140

The objective of a high-resolution study is therefore not only to obtain as small a residual electron density distribution as possible but rather to locate accurately all the atoms and to optimize all experimental parameters. Thus, the normal criteria for the estimation of the accuracy of an X-ray structure determination, such as the standard deviations of the atomic and geometrical parameters, may be not sufficient for judging the accuracy of an EDD determination. The comparison with theoretical calculations or with structure determinations of chemically similar compounds is sometimes necessary.

The experimental and computational effort required for high-resolution studies is much greater than that for a standard X-ray structure determinations. Since thermal displacement of the atoms can lead to a deterioration of the observable features in the EDD, a measuring temperature as low as possible is necessary. A temperature of 100 K, obtained by cooling the sample with a stream of cold nitrogen gas, has proven to be suitable. Recent experiments have shown that it might be worthwhile to lower the temperature even to liquid helium temperature.[31]

It is essential to maintain the temperatures within a small range ($\pm$ 0.5 K) over several weeks and to keep the crystal free of ice. We have developed in our laboratory a cooling device for single crystal diffractometers[29] which allows for this fact. It is also used on the recent new counting systems (CCD counters), although these devices at present time might need some more (software) development in order to be used in this kind of experiment.[32] Furthermore, several sets of symmetry equivalent reflections need to be measured under different experimental conditions in order to eliminate instrumental errors and detect possible double reflections.

A further restriction might be that the crystal must always be of excellent quality. In order to correct for absorption effects it is necessary not only to index all the enclosing faces but also to measure them correctly. If it proves necessary to study the substance in a capillary under an inert gas, then an additional correction for the absorption of the capillary might be useful.[33]

Needless to say that correct absorption and extinction corrections as well as the tracing of unusual atomic displacements (thermal motion) is a prerequisite for obtaining exact deformation densities.

From all this it is obvious that the determination of deformation densities can still not be considered a routine procedure at present. Recommendations for experimental conditions have been made by the IUCr.[34]

Several of the experimental difficulties in EDD determinations as described are considerably reduced when heavy elements are absent. Organic compounds are thus particularly suited for studying developments of the method, and it was for this reason that the Oxalic Acid Project was initiated by the IUCr.[35]

In favorable cases, an effect similar to that of lowering the temperature is achieved when the molecules are held firmly in the crystal lattice by strong intermolecular interactions, such as hydrogen bonds. In such cases, quite acceptable deformation densities have been obtained even at room temperature.

This influence of packing effects is illustrated best by looking at the EDDs of two different classes of organic compounds, one with negligible intermolecular interactions, and one compound which is strongly hydrogen bonded.

As can be seen from Fig. 3, sterically rigid aromatic compounds without dominating intermolecular interactions can be studied without complication using EDD-methods. We undertook a comparative study of 9-*tert*-butylanthracene [1] and its valence isomer as obtained by a photochemical reaction,[24] 9-*tert*-butyl-9,10[Dewar]anthracene [3]. X-ray structure determinations under standard conditions had shown that there was no molecular disorder in the crystals and that the hydrogen atoms could be unequivocally located. Both these factors are among the most important criteria for determining the crystal quality.

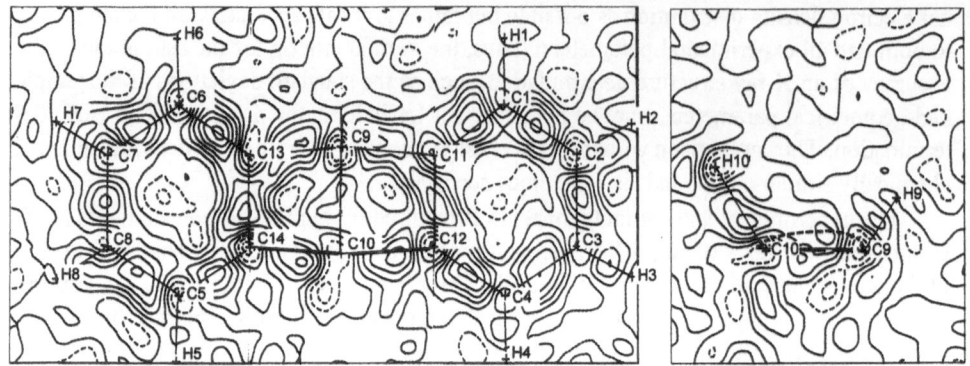

Figure 6. Left: Deformation density in **3**. Contour lines at $0.1\ e/\text{Å}^3$. Right: Longitudinal section through the Dewar bond C9−C1O in **3**. Contour lines at $0.05\ e/\text{Å}^3$. Atoms C9, C10, C11, and H10 lie in the plane of the paper.

Figure 6 shows the deformation density of **3** in a similar orientation to the EDD of **1**. In **1** the EDD maxima lie on axes drawn between the atomic nuclei. The observed EDD is directly proportional to bond parameters obtained by semiempirical methods, such as the bond order (calculated by HMO techniques for the idealized unsubstituted anthracene) and the overlap population (calculated by extended Hückel methods for the idealized 9-*tert*-butylanthracene). Deviations from exact proportionality are only observed in the central ring of the anthracene framework. This is not planar, as was assumed in the calculations, but has a fold angle of 13.8°. This distortion is due to a steric interaction resulting from the large substituent at the 9-position of the anthracene skeleton.

In contrast to **1**, bent bonds are observed in the entire central bicyclic system of the Dewar isomer **3**. Maxima of electron density are not found anymore along the lines joining the nuclei of bonded atoms, but are shifted about 0.2 Å outwards. In addition, the locations of the peak maxima in bonds involving the carbon atom at the 9-position are significantly displaced towards the bridging group. This is similar to the effect which is observed in bonds between atoms with different electronegativities. The Dewar bond (C9−C10 1.623(2) Å) is strongly bent, as can be seen in sections through the bond (Fig. 6 right). The bond length as well as the EDD suggest that this is the weakest bond in the molecule.

The deformation densities given in Fig. 6 can be reproduced by *ab initio* calculations. Very accurate ab initio calculations are necessary in order to reproduce the experimentally observed EDD. Comparison of various calculations have shown that it is essential to use a large set of basis functions o. at least double-ζ quality. In the past it has been found worthwhile to compare experimental with theoretical results. This is especially useful for the detection of effects of intermolecular interactions via the EDD of molecules in a crystal lattice, since these are not usually taken into account in the ab initio calculations.

The analysis of 1,2,3-triazine **4**[36a] serves to illustrate the influence of intermolecular interactions in the crystal lattice on the EDD of the molecule. Figure 7 shows the experimentally determined and Fig. 8 the theoretically calculated deformation density of triazine. The theoretical deformation density of a single molecule has a mirror plane of symmetry passing through C2 and N2. This symmetry is also reflected in the observed bond distances. According to crystallographic criteria,[36b] however, the environment of the molecule in the crystal lattice is asymmetric, this is also clearly seen in the experimentally determined deformation density (Fig. 7). It is immediately apparent that in solid state the EDDs in the C14−H1 and C3−H bonds differ, and the lone-pair electrons at N1 and N3 have different EDDs and spatial distributions.

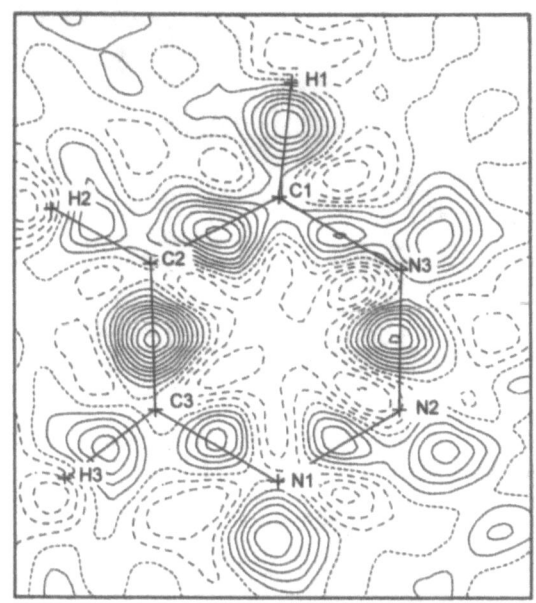

Figure 7. EDD in **4** (mean plane through the N and C atoms). Contour lines at 0.1 $e/\text{Å}^3$.

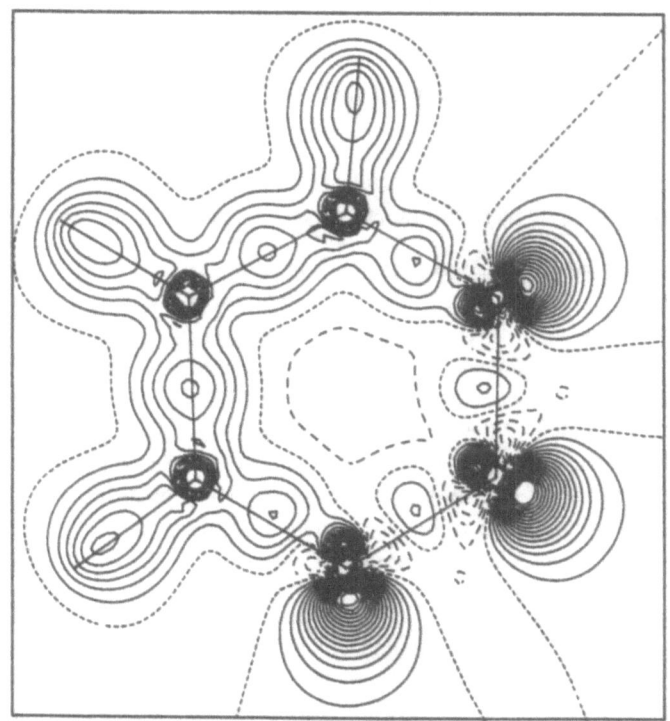

Figure 8. Theoretical EDD of **4**.

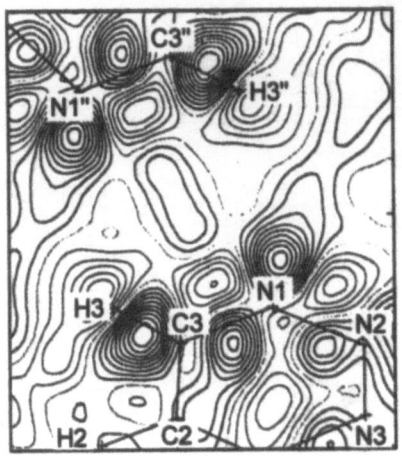 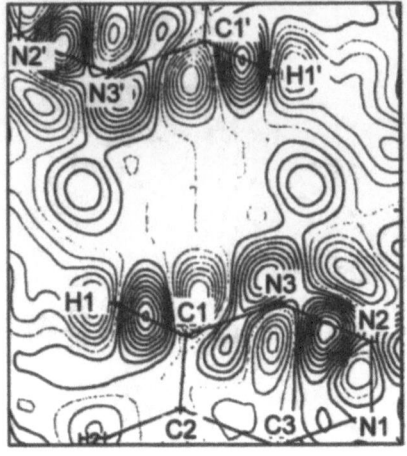

Figure 9   **Figure 10**

The intermolecular EDD in crystalline **4**.

Figure 9: plane through N1, H3, N1", H3". Figure 10: plane through N3, H1, N3', H1'.

Figures 9 and 10 show the EDD in the intermolecular regions where intermolecular interactions occur. The lone pair electrons of N1 and N3 are clearly directed towards H3" and H1' on neighboring molecules. The two intermolecular interactions occur in regions that are crystallographically independent but chemically equivalent. Since it is unlikely that experimental inaccuracies will manifest themselves in a similar manner in different regions of the unit cell, the agreement between the two deformation densities gives a measure of the accuracy of the experiment. The two-dimensional sections are rather unsuitable for the study and display of what are in general three-dimensional intermolecular interactions and of their effect on the intramolecular EDD.

Recent investigations on a series of related heterocyclic compounds, e.g. pyrrole, v–triazole, tetrazole, quinoline, isoquinoline, quinoxaline[37] reveal similar features of the EDD.

The effect observed and described here is not in complete agreement with theoretical calculations on similar systems, which predict that the probability of observing intermolecular interactions might be quite small.[38] However, in numerous studies on organometallic compounds we have obtained also evidence of the influence of packing effects on the EDD of the molecules in a crystal lattice. It was found that in substituted metal carbonyl complexes the lone pairs on the oxygen atoms of the carbonyl groups are polarized in the direction of the neighboring molecules. The differences between experiment and theory indicate that particular attention needs to be paid to intermolecular interactions when interpreting the results of a high-resolution study, even in those cases where such interactions cannot be inferred from the results of a normal X-ray structure determination.

We have tried also to use the method of deformation density determination to detect solid state intermolecular interactions in photoreactive solids.[39] Two prospective candidates were 4-acetoxy-6-[(E)-2-phenylethenyl]2H-pyran-2-one (**5**) and its isomorphous, isoelectronic 4-acetoxy-6-[(E)-2-phenylethenyl]6H-1,3-oxazin-6-one (**6**). The crystal packing of these compounds is illustrated in Figure 11. Irradiation ($\lambda = 360$ nm) of **5** leads to the formation of topochemical [2 + 2] photodimers, whereas **6** is photo-inactive (see Fig. 11), although the distances between the midpoints of the photoreactive C=C double bonds leading to the two possible dimers (Fig. 12) are similar in both cases.

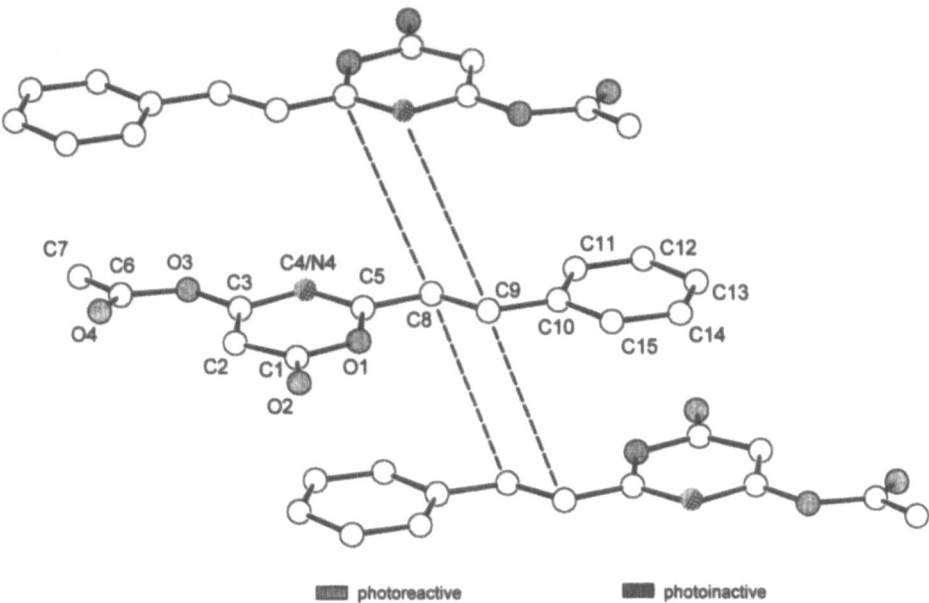

photoreactive   photoinactive

Figure 11. Molecular interactions of **5** and **6** in solid state.

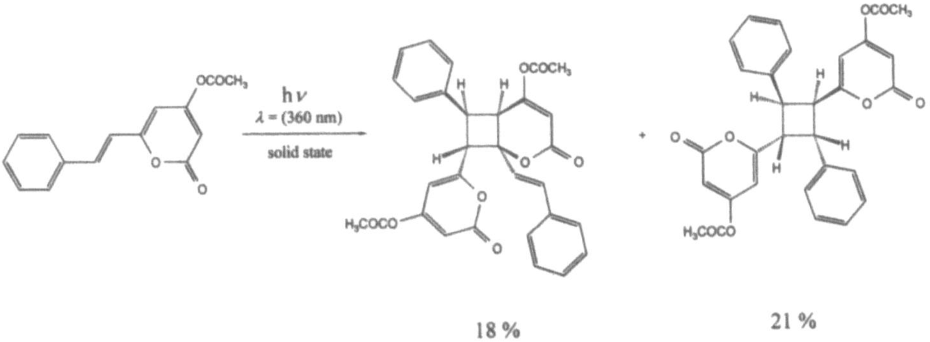

Figure 12. Solid State photodimerisation of **5**.

An explanation for this unusual behavior was sought in the different electron density distributions.

Plots of EED-maps show that the electron density of the photoreactive compound **5** is distorted away from the direction of the interaction so that the major axis of the bonding electron density makes an angle of about 70° with the plane of the molecules. In contrast, the EDD of the photo-inactive compound **6** is as expected for the isolated molecule. The fluorescence spectra of **5** illustrate its photoreactivity in the solid state by monitoring the excimer emission against the isolated molecules in solution. In contrast, compound **6** has a different excited state, which shows the same fluorescence emission in solid state as in solution.[18,40]

The last part of this report is concerned with structures and also solid state reactivities of compounds which are liquids at ambient temperature.

We were interested in the solid state structures of di- and polyenes — also having a prospective photoreactivity in mind — as well as of phosphines. Both classes of compounds are frequently used as ligands in organometallic chemistry, and especially in homogeneous catalysis.

There are modern techniques to crystallize compounds, which are liquids or even gases at room temperature in a cooled capillary *in situ* on the X-ray diffractometer.[41] We obtained single crystals using a micro-zone melting technique with a focused CO_2-laser beam as the heat source.[42] This crystal growing device was adapted for use on a CAD-4 diffractometer,[43] but can be easily transferred to modern CCD-diffractometers. Several crystallization experiments showed that olefins tend to form a supercooled melt or plastic crystals. A similar behavior was observed for bulky phosphines.[43,44] In these cases, difficulty was experienced when trying to initiate the nucleation and to obtain single crystals of high quality. The optimization of the conditions for nucleation and crystal growth revealed that the best nucleation temperature lies approximately 60 K below the melting point of the substance. The ideal temperature for the growth of single crystals is approx. 40 K below the melting point of the compound. However, the quality of the grown crystals is rather limited, therefore it seems that results of EDD-determinations of crystals obtained by these methods have to be taken with great caution.

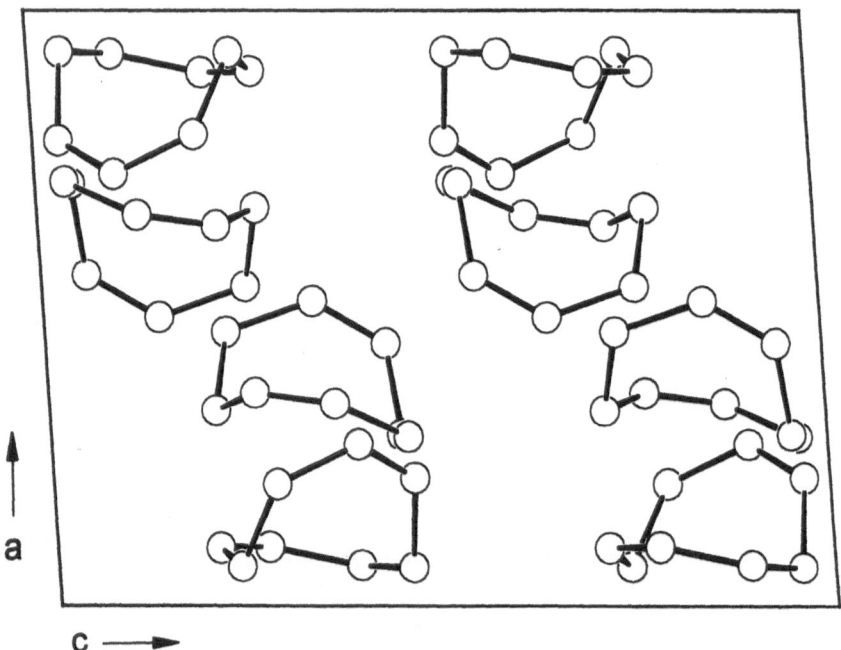

a

c ⟶

Figure 13. The packing of 1,5-COD (**7**).

In Fig. 13, the molecular packing for 1,5-cyclooctadiene (**7**) is given. This compound, a popular ligand in transition metal chemistry, which is readily available from butadiene by cyclo-dimerisation. It crystallizes with two independent molecules in the asymmetric unit of space group $P2_1/c$ in a layer-type structure. For this compound, electron diffraction data,[45] force field geometry optimizations and examination of geometric ring constraints[46] suggested three possible and different distorted ring conformations. These are depicted in Fig. 14 and are to be compared to the results of the X-ray structural investigation.

Single and double bond distances are as to be expected for a non-planar compound with isolated double bonds. In both independent molecules, both types of bonds are equal within the given standard deviations.

146

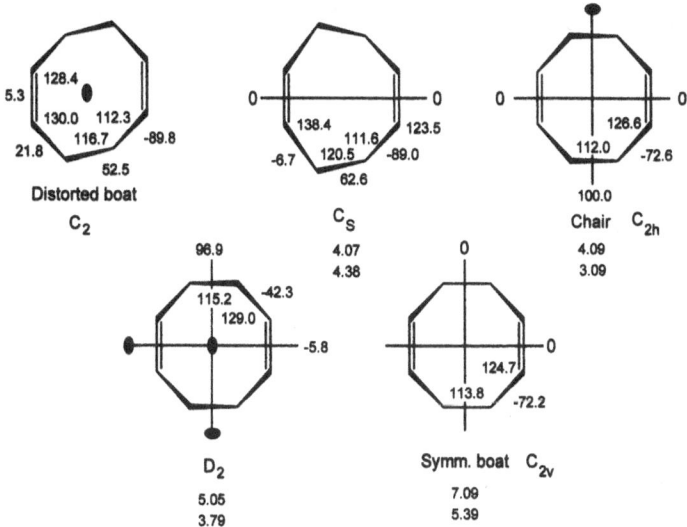

Figure 14. Force field calculated structures of **7**, lower line idealized conformations.[46]

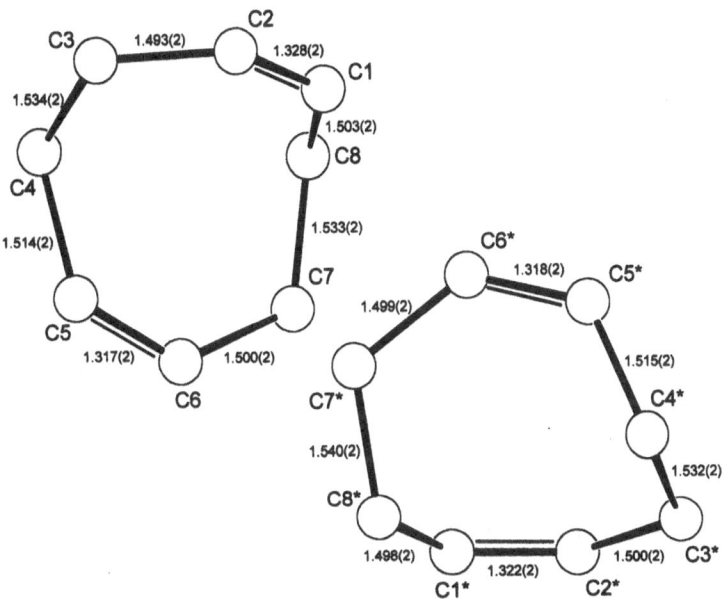

Figure 15. Experimental structure for **7**.

The energetically (6 kJ/mol) preferred conformation, a twist boat form with symmetry *C*2, is found for both independent molecules in the crystal. These differ only by their helicity and in certain conformational details. Most remarkable are close contacts of two hydrogen atoms across the rings (1.99 resp. 2.01 Å). This transanular interaction is in accordance with the observed twofold NMR splitting of the methylene proton singlet into a doublet at 100 K.[47] The deformation densities, however, reveal no unexpected features.

Using these crystal growing techniques we have investigated the solid state structures of several di- and polyenes. These compounds (**8–20**) are summarized in Fig. 16; their crystallographic data are given in Table 2. Several of these structures have been published already,[48] for completeness they are included in this summary.

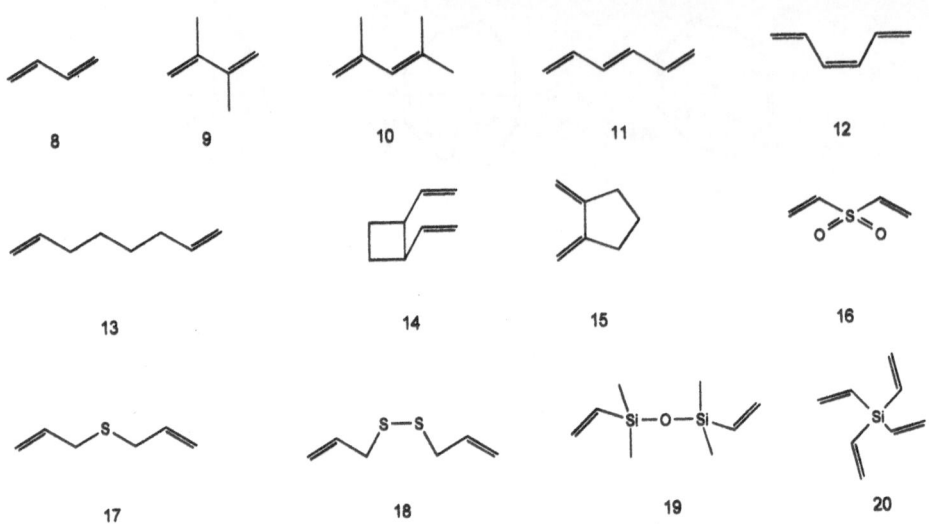

Figure 16. Investigated polyenes as given in Table 2.

Table 2. Crystal data of **8–20**

Compound	Space group	a [Å]	b [Å]	c [Å]	β [°]
1,7-Octadiene	$P2_1/a$	8.710	4.897	9.297	99.500
1,3-Butadiene	$P2_1/n$	6.848	4.169	7.482	109.03
2,3-Dimethyl-1,3-butadiene	$P2_1/n$	7.051	5.582	7.992	113.856
2,4-Dimethyl-1,3-butadiene	$P2_1/c$	6.049	12.310	9.696	106.651
trans-1,3,5-Hexatriene	$P2_1/a$	8.312	4.141	8.361	106.881
cis-1,3,5-Hexatriene	$P2_1/a$	8.325	4.422	8.406	112.540
cis-1,2-Divinylcyclobutane	$P2_12_12_1$	4.180	6.428	25.741	90.000
1,2-Dimethylencyclopentane	$P2_1/c$	7.642	6.153	13.362	103.629
Diallylsulfide	$C2/c$	11.112	10.342	6.219	103.746
Diallyldisulfide	$P2_1/n$	5.346	13.941	10.845	96.152
Tetravinylsilane	$Pbca$	8.995	20.460	15.312	90.000
1,3-Divinyl-1,1,3,3-tetramethyldisiloxane	$P2_1/n$	10.609	7.386	13.280	98.045
Divinylsulfone	$P2_1/a$	10.021	17.713	15.312	91.443
1,1'-Divinylsulfone	$P2_1/c$	14.676	7.434	10.528	102.926

This series of investigations has been started in order to search for possible photochemical solid state [2 + 2] cycloaddition reactions. These investigations were undertaken using a setup for direct irradiation of the single crystals in situ on the diffractometer.[49] As to be expected, the success of these photochemical investigations was limited. This can be readily explained by making use of the packing criteria given by Ramamurthy et al.,[50] which are depicted in Fig. 17.

According to these criteria the corresponding data of **8–20** have been compiled in Table 3. It seems obvious that none of the compounds investigated fulfill these criteria which are a logical expansion of the topochemical postulate.

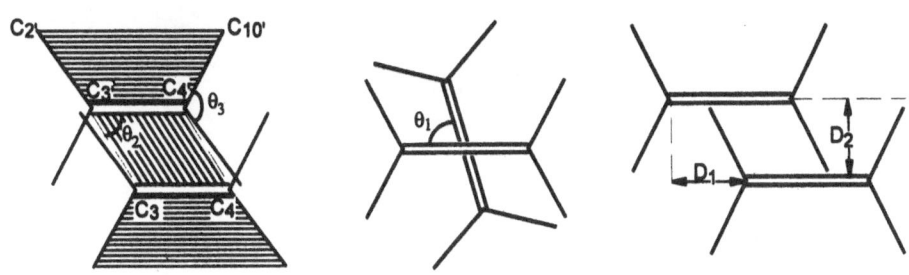

Figure 17. Geometrical parameters used in the relative representation of reactant double bonds.

Table 3. Intermolecular interactions for **8–20** according to Fig. 17

Compound	Distances [Å]	symmetry	$\theta_0[°]$	$\theta_1[°]$	$\theta_2[°]$	$\theta_3[°]$
1,7-Octadiene	4.897	translation	0	0	54.58	99.97
1,3-Butadiene	4.181	translation	0	0	70.38	137.78
2,3-Dimethyl-1,3-butadien	4.972	translation	0	0	67.03	161.22
2,4-Dimethyl-1,3-butadiene	4.352	inversion	0	0	56.88	139.17
trans-1,3,5-Hexatriene	4.148	translation	0	0	69.21	137.12
cis-1,3,5-Hexatriene	4.221	translation	0	0	84.08	136.92
	3.552	inversion	0	0	77.43	119.53
cis-1,2-Divinylcyclobutane	4.180	translation	0	0	80.02	160.84
1,2-Dimethylencyclopentane	35.587	inversion	0	0	62.56	135.72
Diallylsulfide	4.029	inversion	0	0	67.82	136.03
	4.099	inversion	0	0	54.36	102.76
Diallyldisulfide	3.853	inversion	0	0	79.62	123.83
	4.062	inversion	0	0	69.88	130.86
Tetravinylsilane	5.077	translation	0	0	85.53	125.62
	3.996/3.880		23.731	9.560		
1,3-Divinyl-1,1,3,3-	6.762	inversion	0	0	64.83	172.19
tetramethyldisiloxane	3.784/4.310		0	76.691	79.381	
Divinylsulfon	10.022	translation	0	0	63.44	111.73
	3.665/4.098		23.991	22.241		
1,1'-Divinylferrocene	4.462	inversion	0	0	57.58	148.91

These four parameters θ_0, θ_1, θ_2 and θ_3 can be used to describe precisely the relative orientation of reactive double bonds within the unit cell. The parameter θ_0 describes the rotation of the π-system about the double bond axis, θ_1 the deviation of the direction of the double bonds, θ_2 and θ_3 the shift of double bonds along the bond axis and perpendicular to it. The reactivity in solid state is prevented by a large translational shift of the double bonds relative to each other (θ_2 and θ_3) in all compounds under investigation.

Only two solid state reactions have been observed, the dimerisation of *trans*-1,3,5-hexatriene (**11**) according to Fig. 18, and a rapid single-crystal to single-crystal transition of (disordered) *cis*-1,3,5-hexatriene (**12**) into *trans*-1,3,5-hexatriene (**11**) (Fig. 19). This reaction seems to be taking place at the interfaces between the disordered entities.

We have used the experimental setup as described above also to elucidate the molecular structure of several phosphines and di-phosphines.[51] These compounds are used in stoichiometric as well as in transition metal catalyzed reactions as ligands at the transition metal. Small geometric and/or electronic changes in these ligands can cause substantial changes in reactivity or selectivity of these catalysts. We have undertaken a systematic investigation (Fig. 20) of the steric demands of typical phosphine ligands in order to gain more inform-

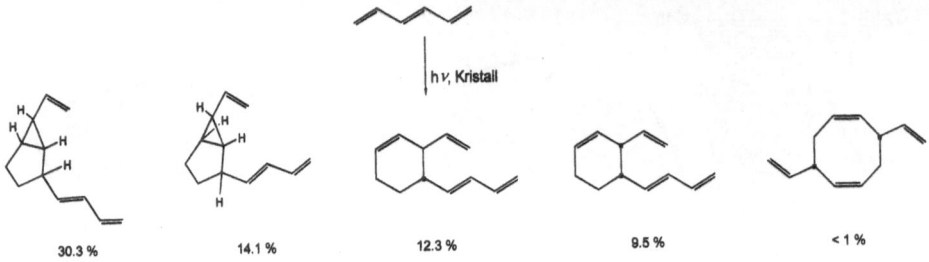

hν, Kristall

30.3 %   14.1 %   12.3 %   9.5 %   < 1 %

Figure 18. Solid state photoproducts of **11**.

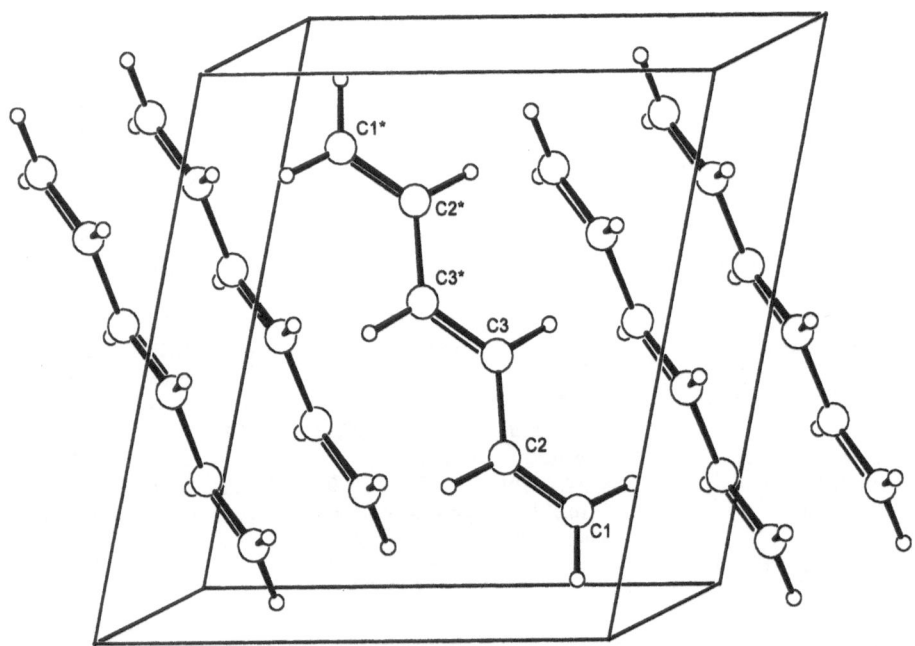

Figure 19. Packing of 1,3,5-hexatriene (**11**).

ation about its influence upon activity and selectivity, and to refine the commonly used cone angle concept of Tolman[52] and its more recent applications and modifications.[53] Some results of these investigations have been published already,[54] and a summary of experimentally determined cone angles is given in Table 4.

Instead of working with only a single conformation we extended Tolman's cone angle model to account for the conformational flexibility of the phosphine ligands. Based on a structural overlay of all possible conformers within a certain energy interval, the size of the accessible molecular surface (ams)[54] is determined by applying Connolly's algorithm.[55] The size of the ams (in $\mathring{A}^2$) for **21–37** is listed in Table 4. In this way we were able to correlate the ams of various catalytically active phosphine-Rh-fragments to their turn-over frequency in the hydrogenation of CO_2 to formic acid, while the original cone angle model and the bite angle concept fail.[54] The ams-model has been proven to be useful in a variety of similar chemical problems.[56]

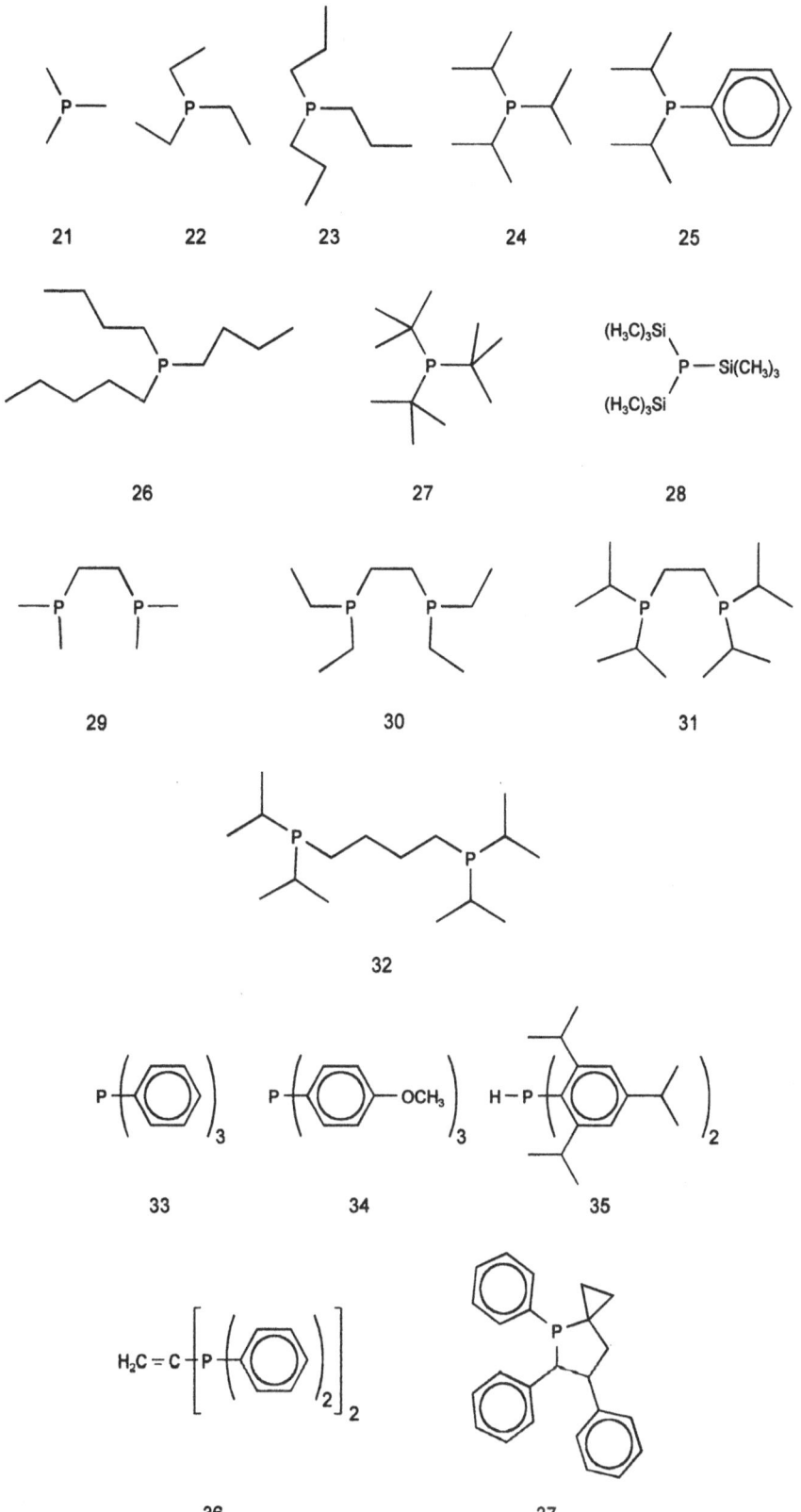

Figure 20. Structural formula for **21–37** (Table 4).

151

Table 4. Geometrical data for phosphines **21-37**

	$P-C(sp^3)$	$-(CH_2)-(CH_2)-$	$-(CH_2)-(CH_3)$	$\Sigma(C-P-C)$	θ	ams
PMe$_3$	1.833(4)	–	–	297.5(3)	113(2)	6.2
PEt$_3$	1.842(2)	–	1.527(3)	298.4(3)	162(2)	3.7
P(n-Pr)$_3$	1.848(2)	1.526(3)	1.520(3)	304.4(3)	144(2)	4.6
P(i-Pr)$_3$	1.861(4)	–	–	309.6(6)	168(2)	2.2
PPh(i-Pr)$_2$	1.857(2)	–	–	307.9(6)	159(2)	2.8
P(n-Bu)$_3$	1.844(2)	1.525(2)	1.515(3)	305.4(3)	159(2)	4.5
P(t-Bu)$_3$	1.911(2)	–	–	322.3(3)	176(2)	1.4
Me$_2$P–(CH$_2$)$_2$–PMe$_2$	1.837(2)	1.529(2)	–	297.3(3)	–	5.6
Et$_2$P–(CH$_2$)$_2$–PEt$_2$	1.847(3)	1.531(3)	1.526(4)	298.5(3)	–	4.0
(i-Pr)$_2$P–(CH$_2$)$_2$–P(i-Pr)$_2$	1.859(2)	–	–	304.2(3)	–	2.9
(i-Pr)$_2$P–(CH$_2$)$_4$–P(i-Pr)$_2$	1.856(3)	-	1.512(3)	304.8(3)	–	3.0

4. CONCLUSION

This contribution has shown that with modern methods of structural chemistry one not only can derive molecular structures of compounds liquid at ambient temperatures, but also due to an increase in the accuracy of diffraction data can draw conclusions about intra- and intermolecular bonding. In certain cases the critical analyses of molecular packing in crystals allows for predictions of solid state reactivity.

Acknowledgement

We wish to thank the Deutsche Forschungsgemeinschaft, the Max-Planck-Gesellschaft, the Fonds der Chemischen Industrie and the BMBF Bonn for financial support and grants.

Note: Structural data of compounds mentioned here which have not yet been published have been deposited in the Cambridge Crystallographic Data File.

REFERENCES

1. H. Tromsdorff, *Ann. Chem. Pharm.* 11:190 (1834).
2. W. Heldt, *Ann. Chem. Pharm.* 63:10 (1847).
3. C. Liebermann, *Chem. Ber.* 10:2177 (1877).
4. C. Liebermann and M. Illinski, *Chem. Ber.* 18:3193 (1885).
5. C. Liebermann, *Chem. Ber.* 28:316 (1895).
6. J. Bertram and R. Kürsten, *J. Prakt. Chem.* 51:316 (1895).
7. H. I. Bernstein and W. C. Quimby, *J. Am. Chem. Soc.* 65:1845 (1943).
8. G. M. J. Schmidt. "Solid State Photochemistry", Verlag Chemie, Weinheim (1976).
9. R. E. Pincock, K. R. Wilson and T. E. Kiovsky, *J. Am. Chem. Soc.* 89:6890 (1967).
10. S. Okamura, K. Hayashi and Y. Kitanishi, *J. Polym. Sci.* 58:925 (1962).
11. P. Yates and P. Singh, *J. Org. Chem.* 34:4052 (1969).
12. C. N. Sukenik, J. A. P. Mandel, N. S. Lau and R. G. Bergman, *J. Am. Chem. Soc.* 97:851 (1977).
13. A. Gavezzotti and M. Simonetta, *Chem. Rev.* 82:1 (1982).
14. V. Ramamurthy and K. Venkatesan, *Chem. Rev.* 87:433 (1987).
15. N. B. Singh, R. J. Singh and N. P. Singh, *Tetrahedron* 50:6441 (1994).
16. F. Toda, *Acc. Chem. Res.* 28:480 (1995).
17. K. Angermund, I. Klopp, C. Krüger and M. Nolte, *Angew. Chem. Int. Ed. Engl.* 30:1345 (1991).
18. I. Ortmann, PhD Thesis, Max-Planck-Institut für Kohlenforschung and University of Wuppertal (1990).
19. P. Debye and P. Scherrer, *Phys. Z.* 19:474 (1918).
20. A. Hartman and F. L. Hirshfeld, *Acta Cryst.* 20:80 (1966).

21. For reviews, see: (a) P. Coppens and M. B. Hall. "Electron Distributions and the Chemical Bond", Plenum Press, New York (1982); (b) P. Coppens, *Angew. Chem. Int. Ed. Engl.* 16:32 (1977); (c) K. Angermund, K. H. Claus, R. Goddard and C. Krüger, *Angew. Chem. Int. Ed. Engl.* 24:237 (1985).

22. (a) M. Breitenstein, H. Dannöhl, H. Meyer, A. Schweig, R. Seeger, U.Seeger and W. Zittlau, *Int. Rev. Phys. Chem.* 3:335 (1983) and references cited therein; (b) R. F. W. Bader and W. H. Henneker, *J. Chem. Phys.* 46:3341 (1967); (c) M. Benard, *J. Am. Chem. Soc.* 100:7740 (1978); (d) F. L. Hirshfeld and S. Rzotkiewicz, *Mol. Phys.* 27:1319 (1974); (e) J. W. Chinn, Jr. and M. B. Hall, *Organometallics* 3:284 (1984); (f) J. W. Chinn, Jr. and M. B. Hall, *J. Am. Chem. Soc.* 105:4930 (1983); (g) P. Coppens and E. D. Stevens, *Adv. Quantum Chem.* 10:1 (1977); (h) K. Hermansson and I. Olovsson, *Theor. Chim. Acta* 64:265 (1984); (i) R. F. Stewart and M. A. Spackman, *in:* "Structure and Bonding in Crystals", vol. 1, M.O. Keefe and A. Navrotsky, eds., Academic Press, New York (1981).

23. For a review, see (a) J. D. Dunitz. "X-Ray Analysis and the Structure of Organic Molecules", Cornell University Press, Ithaca, NY, (1979), and for (b) G. J. H. van Ness and A. Vos, *Acta Cryst.* B34:1947 (1978); (c) J. Lewis, D. Schwarzenbach and H. D. Flack, *Acta Cryst.* A38:733 (1982); (d) K. Hino, Y. Saito and M. Benard, *Acta Cryst.* B37:2164 (1981); (e) A. Mitschler, B. Rees, R. Wiest and M. Benard, *J. Am. Chem. Soc.* 104:7501 (1982); (f) H. Dietrich and C. Scheringer, *Acta Cryst.* B35:1191 (1979); (g) A. Kirfel, G. Will, R. F.Stewart, *Acta Cryst.* B39:175 (1983).

24. (a) B. Jahn, H. Dreeskamp, *Ber. Bunsenges. Phys. Chem.* 88:42 (1984); (b) H. Dreeskamp, B. Jahn and J. Pabst, *Z. Naturforsch.* A 36:665 (1981); (c) K. Angermund, R. Goddard and C. Krüger, *Acta Cryst.* A40:C162 (1984) (Suppl.).

25. F. L. Hirshfeld, *Theor. Chim. Acta* 44:129 (1977).

26. (a) J.-M. Savariault, M. S. Lehmann, *J. Am. Chem. Soc.* 102:1298 (1980); (b) J. D. Dunitz, W. B. Schweizer and P. Seiler, *Helv. Chim. Acta* 66:123 (1983); (c) J. D. Dunitz and P. Seiler, *J. Am. Chem. Soc.* 105:7056 (1983).

27. (a) W.H.E. Schwarz, P. Valtazanos and K. Ruedenberg, *Theor. Chim. Acta* 68:471 (1985); (b) W.H.E. Schwarz, K. Ruedenberg and L. Mensching, *J. Am. Chem. Soc.* 111:6926 (1989); (c) L. Mensching, W. von Niessen, P. Valtazanos, K. Ruedenberg and W. H. E. Schwarz, *J. Am. Chem. Soc.* 111:6933 (1989); (d) W. H. E. Schwarz, L. Mensching, K. Ruedenberg, P. Valtazanos, L. L. Miller and W. von Niessen, *Angew. Chem.* 101:605 (1989).

28. (a) A. Hartman and F. L. Hirshfeld, *Acta Cryst.* 20:80 (1966); (b) A. M. O'Connell, A. I. M. Rae and E. N. Maslen, *Acta Cryst.* 21:208 (1966);

29. (a) P. Coppens, *Science* 158:1577 (1967); (b) P. Coppens, T. M. Sabine, R. G. Delaplane and J. A. Ibers, *Acta Cryst.* B25:2451 (1969); (c) P. Coppens and A. Vos, *Acta Cryst.* B27:146 (1971); (d) D. A. Matthews and G. D. Stucky, *J. Am. Chem. Soc.* 93:5954 (1971); (d) R. Brill, H. Dietrich and H. Dierks, *Acta Cryst.* B27:2003 (1971).

30. (a) J. M. Savariault and M. S. Lehmann *J. Am. Chem. Soc.* 102:1298 (1980); (b) M. A. Spackmann and P. G. Byrom, *Acta Cryst.* B53:553 (1997) and literature quoted therein. For multipolar refinement see also: (c) F. L. Hirshfeld, *Acta Cryst.* B27:769 (1971); (d) R. F. Steward, *J. Chem. Phys.* 51:4569 (1969); (e) R. F. Steward, *J. Chem. Phys.* 57:1664 (1972); (f) R. F. Steward, *J. Chem. Phys.* 58:1668 (1973); (g) R. F. Steward, *Acta Cryst.* A32:565 (1976); (h) R. F. Steward, E. R. Davidson and W. T. Simpson, *J. Chem. Phys.* 42:3175 (1965); (i) R. J. van der Wal and R. F. Steward, *Acta Cryst.* A39:422 (1983); (j) R. J. van der Wal and R. F. Steward, *Acta Cryst.* A40:587 (1984); (k) N. K. Hansen and P. Coppens, *Acta Cryst.* A34:909 (1978); (l) A. S. Brown and M. A. Spackman, *Acta Cryst.* A47:21 (1991); (m) A. Holladay, P. Leung and P. Coppens, *Acta Cryst.* A39:377 (1983); (n) A. Paturle and P. Coppens, *Acta Cryst.* A44:6 (1988); (o) P. Moeckli, D. Schwarzenbach, H.-B. Bürgi, J. Hauser and B. Deley, *Acta Cryst.* B44:636 (1988).

31. (a) G. C. Verschoor and E. Keulen, *Acta Cryst.* B27:134 (1971); (b) K. H. Claus, R. Gerhard and C. Krüger. Martinsrieder Symposium über apparative Entwicklungen in der Röntgen- and Neutronen-Strukturanalyse (1981); (c) D. Zobel, R. Flaig, P. Luger and H.-G. Krane, *Acta Cryst.* A52:C26 (1997).

32. C. Krüger, K. Angermund, J. Bruckmann, F. Lutz and C. Kopiske, *Acta Cryst.* A52:C285 (1996).

33. P. Coppens, *A CA Newsletter* 15:3 (1984).

34. P. J. Becker, P. Coppens and F. L. Hirshfeld, *J. Appl. Crystallogr.* 17:369 (1984).

35. (a) P. Coppens, J. Dam, S. Harkema, D. Feil, R. Feld, M. S. Lehmann, R. Goddard, C. Krüger, E. Hellner, H. Johansen, F. K. Larsen, T. F. Koetzle, R. K. McMullan, E. N. Maslen and E. D. Stevens, *Acta Cryst.* A40:184 (1984); (b) A. A. Pinkerton, C. F. Campana and M. R. Pressprich, *Acta Cryst.* A52:C26 (1997).

36. (a) M. Neunhoeffer, H.-D. Clausen, H. Vtter, H. Ohl, C. Krüger and K. Angermund, *Liebigs Ann. Chem.* 1732 (1985). (b) K. Angermund, PhD Thesis, Max-Planck-Institut für Kohlenforschung and University of Wuppertal (1986).

37. O. Heinemann, PhD Thesis, Max-Planck-Institut für Kohlenforschung and University of Essen (1997).

38. (a) K. Hermansson and S. Lunell, *Acta Cryst.* B38:2563 (1982); (b) I. Olovsson. "Crystal Forces and Hydrogen Bonding. Effect on Charge Density", Laboratoire de Crystallographie, Centre National de la

Recherche Scientifique, Grenoble (1978).

39. I. Klopp, PhD Thesis, Max-Planck-Institut für Kohlenforschung and University of Wuppertal (1990).
40. I. Ortmann, S. Werner, C. Krüger, S. Mohr and K. Schaffner, *J. Am. Chem. Soc.* 114:5048 (1992).
41. (a) R. Boese and D. Blser, *J. Appl. Cryst.* 22:394 (1989); (b) H. Hope, *Acta Cryst.* B44:22 (1988); (c) T. Kottke and D. Stalke, *J. Appl. Cryst.* 26:615 (1993); (d) M. Veith and W. Frank, *Chem. Rev.* 88:81 (1988); (e) M. Renaud and R. Fourme, *Bull. Soc. Fr. Mineral. Cryst.* 89:243 (1966); (f) W. C. von Dohlen and G. B. Carpenter, *Acta Cryst.* 8:646 (1955); (g) D. Brodalla, D. Mootz, R. Boese and W. Osswald, *J. Appl. Cryst.* 18:316 (1985); (h) K. H. Claus and C. Krüger, *Acta Cryst.* C44:1632 (1988).
42. R. Boese and M. Nussbaumer, "*In-situ* Crystallization Techniques", vol. 7 of "IUCR Crystallographic Symposia", Oxford University Press, Oxford, England.
43. F. Lutz, PhD Thesis, Max-Planck-Institut für Kohlenforschung and University of Wuppertal (1992).
44. J. Bruckmann, PhD Thesis, Max-Planck-Institut für Kohlenforschung and University of Bochum 1996.
45. (a) J. D. Dunitz and J. Waser, *J. Am. Chem. Soc.* 94:5645 (1972); (b) N. L. Allinger and T. Sprague, *Tetrahedron* 31:21 (1975).
46. O. Ermer, *Struct. Bonding* 27:161 (1976).
47. F. A. L. Anet and L. Kozerski, *J. Am. Chem. Soc.* 95:3407 (1973).
48. (a) B. Bartkowska and C. Krüger, *Acta Cryst.* C52:1064 (1997); (b) B. Bartkowska and C. Krüger, *Acta Cryst.* C52:1066 (1997).
49. (a) A. Kreuger, *Acta. Cryst.* 8:348 (1955); (b) J. C. Huffmann, W. E. Streib and J. M. Müller, *Proc. Am. Cryst. Assoc.* Ser. 2, 1 156 (1973); (c) A. Simon, H.-J. Deiseroth, E. Westerbeck and B. Hillenkötter, *Z. Anorg. Allg. Chem.* 423:203 (1976); (d) D. Brodalla, D. Mootz, R. Boese and W. Osswald, *J. Appl. Cryst.* 18:316 (1985); (e) MPI für Kohlenforschung, Dept. of Structural Chemistry, unpublished results; (f) R. Boese. Crystal Growing at Low Temperatures, Praktische Aspekte der Kristallstrukturanalyse, 26.8.–28.8.1994, TU Dresden, Abstract.
50. see reference 14.
51. (a) J. Bruckmann, C. Krüger and F. Lutz, *Z. Naturforsch.* 50b:351 (1995) (b) J. Bruckmann and C. Krüger, *Acta Cryst.* C51:1155 (1995); (c) J. Bruckmann and C. Krüger, *Acta Cryst.* C52:1733 (1996); (d) J. Bruckmann and C. Krüger, *Acta Cryst.* C51:1152 (1995); (e) J. Bruckmann and C. Krüger, *J. Organomet. Chem.* 536,537:565 (1997); (f) A. H. Cowley, A. Decken, N. C. Norman, C. Krüger, F. Lutz, H. Jacobsen and T. Ziegler, *J. Am. Chem. Soc.* 119:3389 (1997).
52. (a) C. A. Tolman, *J. Am. Chem. Soc.* 92:2956 (1970); (b) C. A. Tolman, *Chem. Rev.* 77:313 (1977); (c) J. Bruckmann, C. Krüger and F. Lutz, *Z. Naturforsch.* 50b:351 (1995).
53. J. M. Smith, B. C. Taverner and N. J. Coville, *J. Organomet. Chem.* 530:131 (1997) and literature quoted therein.
54. K. Angermund, W. Baumann, E. Dinjus, R. Fornika, H. Görls, M. Kessler, C. Krüger, W. Leitner and F. Lutz, *Chem. Eur. J.* 3:755 (1997).
55. (a) M. L. Connolly, *J. Applied Cryst.* 16:548 (1983); (b) M. L. Connolly, *Science* 221:709 (1983).
56. K. Angermund, S. Geier, P. W. Jolly, M. Kessler, C. Krüger and F. Lutz, in preparation.

METAL AMINE CARBOXYLATES AS HYDRONIUM ION TRAPS. PART 5.1.
THE STRUCTURE OF $K_{1/2}(H_5O_2)_{1/2}\{(-)_D\text{-}trans\text{-}(O_6)\text{-}[Co(1,3\text{-SS-pddadp})]\}\cdot 2H_2O$ (I) DETERMINED AT 18°C AND −100°C AND OF $Li\{(-)_D\text{-}trans\text{-}(O_6)\text{-}[Co(1,3\text{-SS-pddadp})]\}\cdot 7H_2O$ (II) AT 18°C. INTRA- AND INTERMOLECULAR INTERACTIONS IN THE CRYSTALLIZATION OF METAL DIAMINE CARBOXYLATES AND ON HYDRONIUM ION TRAPS

I. Bernal,[a]* James Cetrullo,[a] Jozef Myrczek,[a†] John S. Ricci, Jr.,[b*]
D. J. Radanović[c*] and S. R. Trifunović[c]

[a]Department of Chemistry
University of Houston
Houston, TX 77204-5641
[b]Department of Chemistry
University of South Maine
Portland, Maine
[c]Department of Chemistry, Faculty of Science
Svetozar Markovic University
Kragujevac 3400, Yugoslavia

ABSTRACT

Optically pure (I), $CoK_{0.5}O_{11}N_2C_{13}H_{24.5}$, crystallizes in the orthorhombic system, space group $P2_12_12$, in a cell whose characteristics, at 18°C, are $a = 16.614(2)$, $b = 12.250(2)$ and $c = 9.068(2)$ Å; $V = 1845.4$ Å³, $d(\text{calc}, z = 4, \text{M.W.} = 433.865\text{g mol}^{-1}) = 1.561$ g cm⁻³. At −100°C, the cell constants are: $a = 16.555(2)$, $b = 12.161(2)$, $c = 9.027(3)$ Å; $V = 1817.28$, $d(z = 4, \text{M.W.} = 433.865$ g mol⁻¹$) = 1.694$ g cm⁻³. For the 18°C set, a total of 2564 data were collected in the range $4° \leq 2\theta \leq 55°$ which were corrected for LP factors and absorption effects (psi scans, $\mu = 6.630$ cm⁻¹; rel. trans. coeff. range from 0.9446 to 0.9999); of these, 1562 were unique and had $I \geq 3\sigma(I)$ and were used throughout the calculations described. The absolute configuration was determined from refinements of the solution $(+++)$ and inverted configuration $(---)$. During the test, the $R(F)$ and $R_w(F)$ indices were, respectively, $(+++) = 0.0473$ and 0.0618 and $(---) = 0.0525$ and 0.0644. The final $R(F)$ and $R_w(F)$ values were, respectively, 0.0443 and 0.0527. The same crystal, on the same goniometer head, was used for low temperature data collection, which refined to final $R(F)$ and $R_w(F)$ values of 0.0644 and 0.0756. Coordinates from the absolute determination at 18°C were used as trial coordinates for the low temperature data set, which refined immediately. It is interesting to note that, initially, the salt was prepared as the pure potassium salt but that upon purification (Dowex50 column, chloride form) the salt was isolated as the double potassium-hydronium derivative. The compound, as

---

*Authors to whom correspondence should be addressed.
†On leave from the Technical University of Wroclaw, I-5, Wroclaw, Poland

a racemate, crystallizes as racemic crystals in the monoclinic system ($P2_1/c$) and does not form the hydronium salt (see ref. 20).

Optically pure **(II)**, $LiCoO_{15}N_2C_{13}H_{32}$, crystallizes in the orthorhombic system, space group $P2_12_12_1$, in a cell whose edges are $a = 7.361(3)$, $b = 3.150(2)$ and $c = 22.503(2)$ Å; $V = 2178.18$ Å^3, d(calc., $z = 4$, M.W. $= 522.28$ g mol^{-1}) $= 1.593$ g cm^{-3}. A total of 3015 data were collected in the range $4° \leq 2\theta \leq 56°$ which were corrected for LP factors and absorption effects (psi scans, $\mu = 8.574$ cm^{-1}; rel. trans. coeff. range from 0.9570 to 0.9980); of these, 1628 were unique and had $I \geq 3\sigma(I)$ and were used throughout the calculations described. The absolute configuration was determined from refinements of the solution $(+++)$ and inverted configuration $(---)$. During the test, the $R(F)$ and $R_w(F)$ indices were, respectively, $(+++) = 0.0443$ and 0.0491 and $(---) = 0.0524$ and 0.0553. The final $R(F)$ and $R_w(F)$ values were, respectively, 0.0399 and 0.0453. As a racemate, the compound crystallizes as a racemate in the triclinic space group $P\bar{1}$ (unpublished results, see ref. 21).

We were interested in observing the mode of crystallization as a function of changes in the length of the methylene chains in the amine and carboxylato fragments. In the effect on packing by changing the charge compensating cations (do they cause changes in the mode of crystallization; e.g., racemic to conglomerate?). And, also, in establishing the absolute configuration of the anions in order to correlate this information to the spectrochemical data (CD) of the anions. Upon realizing that they would not crystallize as conglomerates, we were forced to use pre-resolved ligands to prepare the anions.

1. INTRODUCTION

Abbreviations used in this study: edda = ethylenediaminediacetate; edta = ethylenediaminetetraacetate; en = ethylenediamine; ntac = nitriloacetic acid; 1,3-R,R-pddadp = 1-3-diaminopropane-N(R),N'(R)-diacetato-N(R),N'(R)-di-3- propionate; pdta and also trdta = propylenediaminetetraacetate; trien = 1,4,7,10- tetraazadecane.

Note: Conglomerate crystallization is the phenomenon wherein a solution of racemate precipitates enantiomorphic crystals each of which contains exclusively a single enantiomer (homochiral crystals). Naturally, there are crystals of both enantiomorphic types in the crystalline mass, and in studying the conglomerate crystallization of $(-)_{599}$–K[cis-α-Λ(δλδ)-Co(edda)(NO$_2$)$_2$] **(III)**, we[2] concluded that

> we appear "to be witnessing a common origin for the seemingly frequent occurrence of conglomerate crystallization in metal amine carboxylates; namely, they are stereochemically rigid anions with a high tendency to associate in spiral strings which are polymeric in nature and, thus, resemble a polypeptide. These helical strings are formed by (-cation-cation-)$_\infty$ hydrogen bonds which link each cation to two others by three-point attachments reminiscent of the classical three-point attachment postulated as responsible for chiral recognition in biological processes; for example, as in enzymatic catalysis. In a racemic solution both types of spirals are formed since a racemate is equally composed of L and D moieties; however, since many crystallize as enantiomorphic crystals, the obvious conclusion is that the stitching together of the spiral strings, by the counter ions and the waters (if present), must lead to better energies of crystallization when they segregate into homochiral lattices during conglomerate crystallization."

Our reference to a common pattern of crystallization among metal derivatives of amine carboxylates is due to the observation that the crystal structures of the following compounds are known to be conglomerates:

$NH_4[Co(edta)]\cdot2H_2O$ **(II)**[3], $K[Co(edta)]\cdot2H_2O$ **(III)**[3], and $Rb[Co(edta)]\cdot2H_2O$ **(IV)**[3] (space group $P2_12_12_1$ for all three); $K[Mn(edta)]\cdot2H_2O$ **(V)**[4] (space group $P2_12_12_1$); $K[Al(edta)]\cdot2H_2O$ **(VI)**[5] (spacegroup $P2_12_12_1$); $Na_2[H_2OMg(edta)]\cdot5H_2O$ **(VII)**[6] (space group C2); K[Co(pdta)]·2H$_2$O (**VIII**)[7] (space group B22$_1$2); Na$_2$[(edta)W(μ_2-OH)$_2$W(edta)] (**IX**)[8] (space group P2$_1$); Rb$_2$(edta)·2H$_2$O (**X**)[9] (space group P2$_1$); [Co(NH$_3$)$_6$][Na(edta)]$_3$·5H$_2$O (**XI**)[10] (space group P2$_1$; [Sn$_2$(edta)]$_n$·2nH$_2$O (**XII**)[11] (space group P1); the sodium Na[Cu(ntac)]·H$_2$O (**XIII**)[12] (space group P2$_1$2$_1$2$_1$), and lithium salt Li[Ca(ntac)]·3H$_2$O (**XIV**)[13] (space group P2$_1$2$_1$2$_1$); Na[Ca(ntac)]·H$_2$O (**XV**)[14] (space group P2$_1$3); Na[M(trdta)]·3H$_2$O with M = V[15] (**XVI**), Cr[16] (**XVII**) and Rh[16] (**XVIII**), all three space group P2$_1$2$_1$2$_1$; Na$_2$C[Mn(edta)]·5H$_2$O (**XIX**)[17] crystallizes in the space group C2. Finally, the salt K[Mn(trdta)]·3H$_2$O also crystallizes in space group P2$_1$2$_1$2$_1$.

Since in every case, the optically active crystals used in the structural study were the result of conglomerate crystallization, such large number of "accidents" suggest that these amine-carboxylates have a marked tendency to crystallize by such pathway and that, if systematic studies of this phenomenon were to be carried out, one may find an even larger number of cases of conglomerate crystallization among compounds of this class. The important question is: which amine carboxylates are prone to crystallize as conglomerates; and, hopefully, why?

It is notable that the Cambridge Structural Database (CSD)[19] contains 44 structures of Co(III) diaminetetracarboxylates of which 24 were reported to crystallize in enantio-morphic space groups; and, of these, only four are cases of conglomerate crystallization; e.g., M[Co(edta)·2H$_2$O, M = NH$_4$, K and Rb[3] and that of K[Co(pdta)]·2H$_2$O (**VIII**)[7] (space group B22$_1$2). However, this is about 9 % of the total which is about twice the frequency of the occurrence of conglomerate crystallization for the entire CSD.

In this report, we note that (a) the spiral string formation we noted[2] in the reported packing of NH$_4$[Co(edta)]·2H$_2$O (**II**)[3], K[Co(edta)]·2H$_2$O (**III**)[3], Rb[Co(edta)]·2H$_2$O (**IV**[3], K[Mn(edta)]·2H$_2$O (**V**)[4], K[Al(edta)]·2H$_2$O (**VI**)[5], K[Co(pdta)]·2H$_2$O (**VIII**)[7] and of (−)$_{599}$-K[cis-α-Λ(δλδ)-Co(edda)(NO$_2$)$_2$] (**III**),[1] occurs in spite of the changes in charge compensat-ing cations in (**II**), (**III**) and (**IV**), (b) in spite of the change of an octahedral [Co(III)] cation to a Jahn-Teller distorted [Mn(III)] or (c) to a group IIIA element [Al(III)]. As in the case of (**III**), conglomerate crystallization also occurs in (**VIII**) despite the change of the central amine from a 1,2-diaminoethane to a 1,3-diaminopropane. We were, thus, curious to invest-igate what changes in charge compensating cation and what changes in the methylene chain sizes would result in changes in the crystallization pathway.

As a result of the above facts, we decided to explore changes in both the cations and the length of the methylene chains by establishing the crystallization mode of (**I**) and (**II**); particu-larly, in view of the close resemblance of the amine carboxylate ligand present therein and that in (**VIII**) (a 1,3-diaminotetraacetate) which, as the potassium salt, crystallizes as a conglom-erate. Initial results with racemic solutions demonstrated that, unlike the above cobalt species quoted above, K[Co(1,3-SS-pddadp)]·2H$_2$O crystallizes as a racemate, space group P2$_1$/c.[20] We determined that racemic solutions of its lithium salt (**II**) also crystallize as a racemate (for space group and cell constants see ref. 21). Therefore, since the absolute configurations were needed to be established for further chiroptical studies being carried at one of our laboratories (D. J. R.) we established them here, and that is the subject of this report.

2. EXPERIMENTAL

Syntheses: The synthetic and resolution procedures are detailed in a paper by D. J. Radanović et al.[22], where elemental analyses are given and a variety of spectral data are also detailed.

Note that purification of the compound is stated[22] to be through a Dowex 1-8X(200-400 mesh) anion exchange resin column in the chloride form which was washed with water and eluted with 0.1 M KCl. Two bands were obtained which were desalted by passage through a G-10 Sephadex column with distilled water as the eluant. The first band (blue) was that of

K(*trans* O56)[Co-(1,3-pddadp)]·2H$_2$O whose elemental analysis was given and which corresponds to this formula and not that of (**I**). For details of the separation, purification, reconversion of the resolved anion to the potassium and lithium salts, and their elemental analyses, the reader is referred to the original.[22]

X-Ray Data Collection: The crystals used in the structure determination were selected randomly, on the basis of their sizes and shapes, and glued onto the glass fiber of a goniometer head. Data were collected with an Enraf-Nonius CAD-4 diffractometer operating with a Molecular Structure Corporation TEXRAY-230 modification[23] of the SDP-Plus software package.[24] The crystals were centered with data in the $18° \leq 2\theta \leq 32°$ range and examination of the cell constants and Niggli matrix[25] showed both to crystallize in a primitive, orthorhombic lattice whose systematic absences belong, in the case of (**I**) to 2_12_12, and in the case of (**II**) to the space group $P2_12_12_1$. The intensity data sets were corrected for absorption using empirical curves derived from Psi scans[23,24] of six reflections, each. The scattering curves were taken from Cromer and Waber's compilation.[26] The low temperature data set for (**I**) was collected using the same crystal as the 18°C set.

The structures were solved from their Patterson maps, using the Co atom as the heavy atom. After refinement of the scale factor and the positional parameters of the Co atoms, a difference Fourier map produced most of the non-hydrogen atoms; the rest were found in subsequent difference maps. Heavy atoms were refined anisotropically until convergence, at which point the hydrogen atoms of the H$_5$O$_2^+$ and those of most of the waters were found in difference maps and added; the hydrogen atoms of the anions were added at idealized (C−H = N−H = 0.95 Å) positions. All non-hydrogen atoms were, again, refined anisotropically until convergence. At the end of the refinement, the values of the $R(F)$ and $R_w(F)$ factors were computed, till convergence, for the two possible enantiomorphic configurations of the Co anions. The results were as follows: For (**I**; 18°C), $R(F)$ and $R_w(F)$ for the original $(+++)$ coordinates were 0.0473 and 0.0618, respectively; for its enantiomorph $(---)$ the values were 0.0525 and 0.0644. For (**II**), $R(F)$ and $R_w(F)$ for the original $(+++)$ coordinates were 0.0443 and 0.0491, respectively; for its enantiomorph $(---)$ the values were 0.0524 and 0.0553. Thus, the coordinates were appropriately set, and this defined the absolute configuration of the Co anions for our data collection crystals. Final $R(F)$ and $R_w(F)$ values for (**I**) and (**II**), were, respectively, 0.0443 and 0.0527 and 0.0399 and 0.0453. Details of data collection and processing for the 18°C data sets for (**I**) and (**II**) are given in tables deposited with the Cambridge Structural Database. respectively. Finally, since the absolute configuration for (**I**) was already known from the 18°C data set, and we used the same the crystal for collecting the low temperature data, the final coordinates from the former set were used as the start of the refinement for the latter set, which refined to $R(F) = 0.0644$ and $Rw(F) = 0.0756$. From now on, we will limit discussion of the stereochemistry of (**I**) to values derived from the low temperature data. Information obtained with (**I**) at 18°C is effectively identical with that obtained at −100°C and it is, thus, redundant.

Figure 1 gives a labeled view of the cation H$_5$O$_2^+$ present in (**I**). Figure 2 shows the anion present in (**I**) and its environment. Figure 3 gives a labeled view of the anion in (**II**) and its environment. In all cases the anions are depicted in their correct absolute configuration for the chiroptical symbol appearing on the title. Final values of the fractional coordinates with equivalent-isotropic thermal parameters for (**I**) and (**II**) as well as bond lengths, bond angles and selected hydrogen bonds for (**I**) and (**II**) have been deposited with the Cambridge Structural Database. Anisotropic thermal parameters, torsional angles and structure factors for (**I**) and (**II**) are submitted as Supplementary Material.

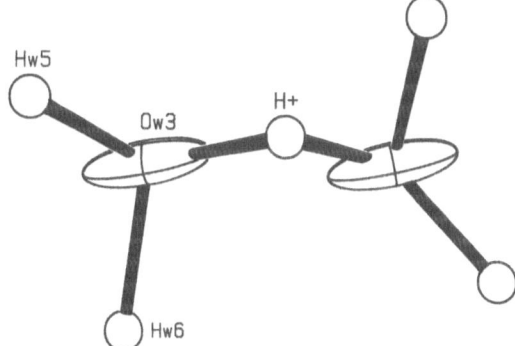

Figure 1. A labeled representation of the stereochemistry of the $H_5O_5^+$ cation in compound (**I**). Note the ellipsoids of thermal motion of the oxygens, the major axis of the ellipsoid suggesting large amplitudes of motion to, and away, from the central proton. We believe this is caused by the partial disorder of the K^+ cations which, when occupying the position of the proton, are bound to move the oxygens away from the positions they would normally occupy. Note that, with the exception of the hydrogens of Ow2, the hydrogens of the remaining waters of hydration were found at chemically sensible positions.

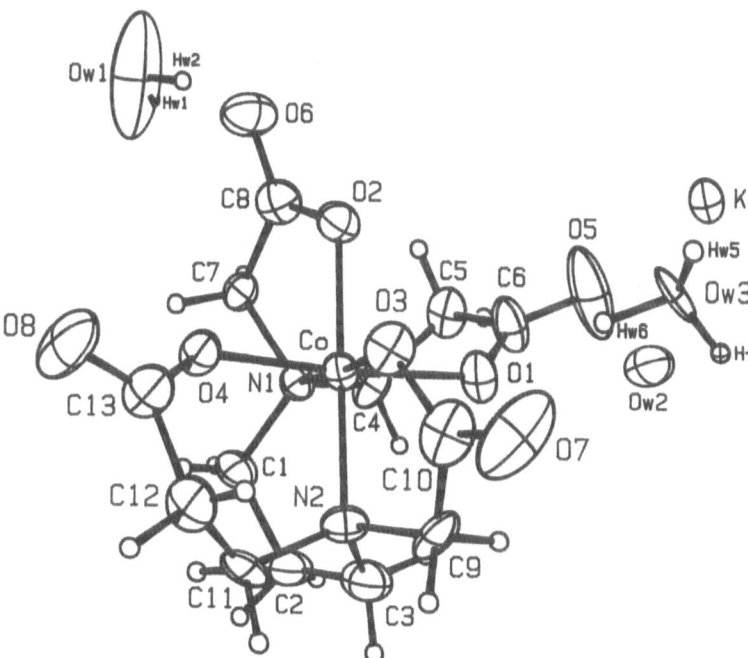

Figure 2. A view of the asymmetric unit of compound (**I**). Note that Hw2, of Ow1, is positioned at the correct place to form a hydrogen bond with O6 , while Hw6 of Ow3 is positioned at the correct location to hydrogen bond with O5.

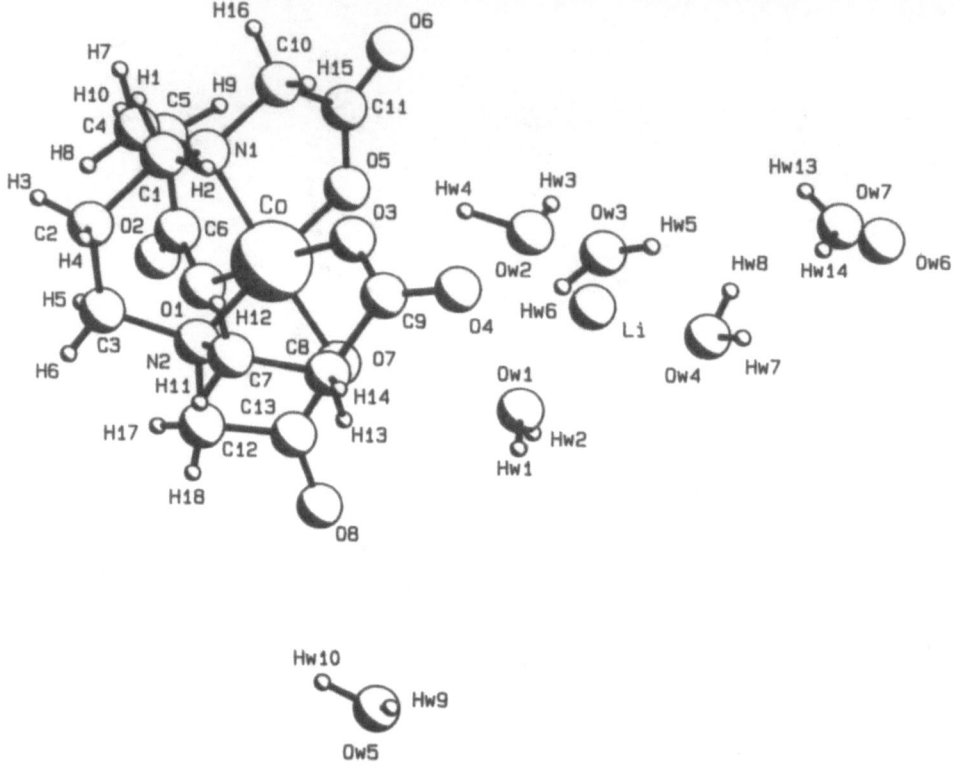

Figure 3. A labeled representation of the stereochemistry of the anion, cation and waters of crystallization present in the asymmetric unit of (**II**). Note that, with the exception of the hydrogens of Ow6, the hyd;rogens of the remaining waters of hydration were found at chemically sensible positions.

3. RESULTS AND DISCUSSION

The Nature of the Compounds (I) and (II), As Determined Crystallographically

The structure of (**II**) is completely non-controversial in that the contents of the asymmetric unit are exactly of the composition of the crystals as analyzed immediately after its synthesis and after its optical resolution.[22] However, the nature of (**I**) was a total mystery to us when we first determined its structure at room temperature, and that is why we redetermined it at low temperature in order to increase resolution in the region of the hydronium ion.

As long ago as 1992, for us it was a great surprise that the two salts differed in the nature of their charge compensating cations since they had a common origin, and the crystals had been grown from deionized water. At the time we had no explanation for these results and set them aside to see if we could produce a sensible account for the change in composition for the potassium salt, but not for that of the lithium one.

One possibility that occurred to us was that the solution of the potassium salt had, accidentally, been acidified. However, this was not only not the case, and we also were aware of the fact that despite passing a Co(edta)⁻ solution through an acid Dowex50 column (in a solution of $pH < 1.0$), the isolated species was an [aquoCo(Hedta)]·3H$_2$O compound with one dangling, protonated, acetate chain.[27] There was no hydronium ion trapped in this case. Thus, we were at an impasse at this juncture. However, for simple reasons of charge balance, there had to be a hydronium ion present in the lattice and, since it was very similar in structure to the classically known $H_5O_2^+$ cation, there was nothing controversial as to its nature, only as to its origin.

Table 1.

Compound Isolated	Space Group	Conglomerate	Reference
$(H_3O^+)[Co(en)_2(ox)]Cl_2 \cdot 2H_2O(1)$	$P2_1/c$	No	1(a)
$(H_5O_2^+)[cis\text{-}\alpha\text{-}Co(edda)(ox)](2)$	$P2_1/n$	No	1(b)
$(H_5O_2^+)[trans\text{-}Co(\beta\text{-}ala)_2(NO_2)_2](3)$	$P2_1/n$	No	1(c)
$K[cis\text{-}\beta\text{-}Co(edda)(ox)] \cdot H_2O(4)$	$P2_1/c$	No	1(b)
$K[cis\text{-}\alpha\text{-}[Co(edda)(malonate)] \cdot 3H_2O(5)$	$P2_1/n$	No	1(b)
$K(H_9O_4^+)[cis\text{-}(NO_2)_2Co(gly)_2](6)$	$P2_1/n$	No	1(d)
$[Co(Hedta)(H_2O)] \cdot 3H_2O(7)$	$P2_12_12$	Yes	27
$K[Co(pdta(H_2O)] \cdot 2H_2O(8)$	$B22_12$	Yes	7
Compound (I)	$P2_12_12$	No[a]	This Study
Compound (II)	$P2_12_12_1$	No[a]	This Study

[a]Prepared as pure chiral anions by the use of resolved ligands. Their racemic salts crystallize from deionized water as racemates, see refs. 20 and 21.

The Nature of the Hydronium Ion in (I)

To begin with, let us say we suspect that the ease with which we located the central hydrogen of the $H_5O_2^+$ cation was due to (a) for charge and symmetry reasons, the proton must lie on a two-fold axis, in between two waters related by the self-same two-fold axis, see Fig. 1. In fact, the crystals are sufficiently high quality that the water hydrogens were readily located, experimentally and (b) there is probably a small amount of disorder between H^+ and K^+, giving rise to a somewhat larger electron density at the former site than one would normally expect for half of a hydrogen ion. This result became far clearer in the low temperature data set and was one of the prime reasons for us collecting both. Hereafter, we will use arguments based on the information obtained with the low temperature data set.

The Solution of the Mystery of the Presence of the Hydronium Ion in (I) by the Results of Later Studies of Co(III) Amine Carboxylates

The fact that we isolated a hydronium potassium double salt from a solution which, initially, contained a pure potassium salt was something of a mystery to us for several years, until we begun noticing additional cases of hydronium ion trapping in the case of other amine carboxylates exhibiting the following, fascinating results:

(a) minor variations in composition and/or stereochemistry of the cobalt amine carboxylates is enough to influence the outcome. Such a statement is based on the following experimental observations detailed in Table 1.

(b) note, also, that despite passing a Co(edta) solution through an acid Dowex50 column, in a solution of $pH < 1.0$, the isolated species was compound (6, in Table 1) with one dangling acetate chain.[27] There was no hydronium ion trapped, instead one carboxylate chain was protonated, no longer bonded to Co(III), and replaced by a water ligand. Therefore, the fact that the solution is acidic does not guarantee the trapping of hydronium ions even if present in the solution. Therefore, the amine carboxylate anions has to have well defined stereochemical characteristics.

(c) crystals of compounds (1) and (2) (Table 1) were grown from slightly acidified solutions; thus, initially, we thought that was the origin of the hydronium ions trapped. However, those of (3) and (6) were grown from deionized water solutions of their potassium salts and the first batches of crystals to come out of solution (for details, see original references 1(b), 1(c) and 1(d)) are hydronium ion salts of varying composition. This result is reproducible and it is documented with elemental analyses, see refs. 1(a) to 1(d). In the case of (3), after several batches of crystals were isolated as the $H_5O_2^+$ salt, we obtained crystals of the pure potassium salt and determined its crystal structure. For details, see ref. 1(c).

(d) comparison of compounds (2) and (4) show that changing the configuration of the Co anion from *cis-α* to *cis-β* is sufficient to change the outcome of the crystallization pathway. The first batches of crystals of the former are isolated as the pure $H_5O_2^+$ salt, while the latter always produces the pure potassium salt. Thus, elemental composition alone is not sufficient to insure trapping of hydronium ions as the geometric change herein described clearly shows.

(e) under the same crystallization conditions, our Li salt (**II**) also does not produce a hydronium containing species despite the fact it contains the exact same anion. Thus, the nature of the charge compensating cation appears to influence the outcome as well, and only potassium salts have, thus far, produced hydronium ion salts or mixed potassium-hydronium salts. We expect to explore this observation in more detail to ascertain if the statement is incorrect; or, if correct, why this is the case.

(f) it is interesting to note that Co(III) amine carboxylates are able to, selectively, trap hydronium ions of three different kinds. So far we have not been able to trap $H_7O_3^+$ salts; however, the $H_9O_4^+$ salt is unique in that it is a polymeric, infinite, "hydrogen ion wire". For details see ref. 1(d).

To conclude this section, the isolation of (**I**) from its pure potassium salt was a mystery before; however, its origin is now clear as documented here. One important point to bring forth is that the concentration of H_3O^+ in deionized water cannot exceed 10^{-7} moles liter^{-1}, whereas the concentrations of the potassium salts is at least 0.1 to 0.2 molar, the implication being that the solubilities of the isolated hydronium ion salts must be markedly smaller than those of the potassium counterparts. Finally, in at least one case, we sequentially isolated both species from the same solution. The order in which they were isolated clearly indicates that the potassium salt is, by far, the most soluble one.

The Mode of Crystallization of Compunds (I) and (II)

Before proceeding with the discussion the crystallization mode, we must bring forth an important issue concerning the chiro-optical symbol for the anions present in (**I**) and (**II**); namely, note that although the absolute configuration at the nitrogens of the unbound ligand, used to prepare both compounds, was known to be (R)[22], the fact of the matter is that, upon binding to cobalt, the absolute configurational symbol at the nitrogens is reversed to (S). The reason for this change is one of the weaknesses of the Cahn, Ingold and Prelog[28] notation since the lowest ranking point for the unbound ligand nitrogens is the non-bonded electron pair. Upon binding to the metal, this point is occupied by cobalt, and becomes the highest ranking one; therefore, the correct notation for the chiroptical symbol at nitrogens, in the cobaltate anion, is (S). Thus, the correct symbols for the anions present in (**I**) is given by the chiroptical symbol $K_{1/2}(H_5O_2)_{1/2}\{(-)_D$-*trans*-$(O_6)$-[Co(1,3-SS-pddadp)]$\}\cdot 2H_2O$ while that for (**II**) is Li$\{(-)_D$-*trans*-(O_6)-[Co(1,3-SS-pddadp)]$\cdot 7H_2O$.

Now, returning to the mode of crystallization of (**I**) and (**II**): part of the reason for carrying out this study, as mentioned in the Introduction, was the desire to document what changes in composition and/or stereochemistry of Co(III) amine carboxylates were responsible for the selection of crystallization mode (e.g., racemic vs. conglomerate). Reference to compounds mentioned in the Introduction and in Table 1 provide some interesting glimpses into this phenomenon; to wit:

If we limit ourselves strictly to Co(III) compounds, then $NH_4[Co(edta)]\cdot 2H_2O$ **(II)**[3], $K[Co(edta)]\cdot 2H_2O$ **(III)**[3], $Rb[Co(edta)]\cdot 2H_2O$ **(IV)**[3] crystallize as conglomerates and are isomorphous and isostructural with one another. This is not surprising since many are the examples of $NH_4{}^+$, K^+ and Rb^+ compounds that are isomorphous and isostructural; for example, the alums. The amine carboxylate ligand (edta) forms five-membered rings when bound to a metal. $(-)_{599}$-$K[cis$-α-$\Lambda(\delta\lambda\delta)$-$Co(edda)(NO_2)_2]]$ **(III)**,[1] also crystallizes as a conglomerate. Its amine carboxylate ligand forms three five-membered rings having lost two acetate chains which were replaced by two $-NO_2$ ligands — a significant change in composition and in stereochemistry, but one that preserves the ability of the anions to form hydrogen bonded, infinite helical strings, linked by potassium ion (for arguments, see ref. 1). $K[Co(pdta)]\cdot 2H_2O$ **(VIII)**[7] has four acetato chains and an interesting change in the diamine portion of the ligand; namely it contains an 1,3-diamino propane which give rise to four five-membered rings from the acetato moieties and one six-membered ring from the diamine. It crystallizes as a conglomerate also — all compounds mentioned thus far crystallize in the space group $P2_12_12_1$. However, when we come to $M[Co(1,3$-pddadp)$\cdot XH_2O^{20\text{-}22}$ ($M = K$, Li) the crystals we obtained were racemic. Thus the substitution of two propionate chains for two acetate chains is enough to change crystallization mode. The obvious conclusion is that the introduction of one additional $-CH_2-$ to the trans-axial carboxylato chains forces a change in the crystallization pathway. Why? Because it changes the stereochemistry of the oxygens needed to form the hydrogen bonds that create the helical strings. We have noted this same phenomenon previously, when comparing the crystallization behavior of $[cis$-$Co(en)_2ox]^+$ vs. $[cis$-$Co(pn)_2ox]^+$ cations in a variety of salts. The former crystallized as conglomerates, the latter as racemates.[29-30]

The Conformation of the Anions in (I) and (II)

The oxygen ligands of the propionate chains occupy the trans-axial positions of the anion, as shown in Figs. 2 and 3, whereas the acetato oxygen ligands occupy the more strained[3] equatorial positions. This is true for both compounds and it is a surprising result since, normally, these species should prefer to minimize steric strain. The effect of such steric strain on physical properties of diamine tetracarboxylates of the transition metal ions has been the subject of extensive studies, particularly of the CD and nmr changes introduced into these hexadentate anions upon changing either the chain lengths or the relative orientation of the hetero chain species; e.g., O5O6 vs. O6, etc.[20,22,31-33]

Acknowledgment

We thank the Robert A. Welch Foundation for support of this research (Grant E-592 to I. Bernal) and for fellowships granted to J. Cetrullo and J. Myrczek. We also thank the US National Science Foundation for the funds used in purchasing the diffractometer (at UH), and for grant 8818818 to B. E. Douglas and D. J. Radanović.

REFERENCES

1. (a) $(H_3O^+)[Co(en)_2oxolato]C_{12}\cdot H_2O$, I. Bernal, J. Cai and J. Myrczek, *Polyhedron*, 12:1157 (1993); (b) $(H_5O_2^+)[cis$-α-$Co(edda)ox]$, I. Bernal, J. Cai and W. T. Jordan, *J. Coordinat. Chem.* 37:283 (1996); (c) $(H_5O_2^+)[trans$-$Co(\beta$-alaninato)$_2(NO_2)_2$, I. Bernal, J. Cai, F. Somoza, and S. S. Massoud, *Inorg. Chim. Acta* in press (1997); (d) $(H_9O_4^+)[cis$-dinitro-$Co(trans$-N-N-glycinato)$_2]_2$, F. Somoza, J. Cai, and I. Bernal, *J. Coord. Chem.* in press (1997).
2. I. Bernal, J. Cetrullo, J. Myrczek, J. Cai and W. T. Jordan, *J. Chem. Soc. DALTON*, 1771 (1993).
3. H. A. Weakliem and J. L. Hoard, *J. Amer. Chem. Soc.* 81:549 (1959).
4. (a) T. Lis, *Acta Crystal.* B34:1342 (1978). (b) J. Stein, and J. P. Fackler, Jr., *Inorg. Chem.* 18:3511 (1979).

5. T. N. Poljanova, N. P. Bel'skaya, D. Tyurk de De Garcia Baños, M. A. Porai-Koshits, and L. I. Martynenko, *J. Struct. Chem.* 11:158 (1970).

6. J. J. Stezowski, R. Countryman, and J. L. Hoard, *Inorg. Chem.* 12:1749 (1973).

7. R. Nagao, F. Marumo, and Y. Saito, *Acta Crystal.* B28:1852 (1972).

8. S. Ikaki, Y. Sasaki, A. Nagasawa, C. Kabuto, and T. Ito, *Inorg. Chem.* 28:1248 (1989).

9. M. Cotrait, *Acta Crystal.* B26:107 (1970).

10. E. O. Schlemper, *J. Cryst. Mol. Struct.* 7:81 (1977).

11. F. P. van Remortere, J. J. Flynn, F. P. Boer, and P. P. North, *Inorg. Chem.* 10:1511 (1971).

12. S. H. Whitlow, *Inorg. Chem.* 12:2286 (1973).

13. V. V. Fomenko, L. I. Kopaneva, M. A. Porai-Koshits, and T. N. Polyanova, *J. Struct. Chem* 25:344 (1974).

14. B. L. Barnett and V. A. Uchtman, *Inorg. Chem.* 18:2674 (1979).

15. J. C. Robles, Y. Atsuzaka, S.Inomata, M. Shimoi, W. Mori, and H. Ogino, *Inorg. Chem.* 32:13 (1993).

16. R. Herak, G. Srdanov, M. Djuran, D. J. Radanović and M. Bruvo, *Inorg. Chim. Acta* 83:55 (1984).

17. X. Solans, S. Gali, M. Font-Altaba, J. Oliva, and J. Herrera, 1988, *Afinidad* 45:243 (1988).

18. Private communication by Professor H. Ogino, Tohoku University, whom we thank. This study is not in the Cambridge Structural Database (ref. 19), release of April, (1997).

19. Cambridge Structural Database: F. H. Allen, J. E. Davies, J. J. Galloy, O. Johnson. O. Kennard, C. F. Macrae, E. M. Mitchell, G. F. Mitchell, J. M. Smith, and D. G. Watson, 1991, *J. Chem. Info. Comp. Sci.* 31:187. Version released April (1997).

20. M. Parvez, C. Mariconbdi, D. J. Radanovic, S. R. Trifunović, V. D. Miletić and B. E. Douglas, *Inorg. Chim. Acta* 248:89 (1996).

21. I. Bernal, unpublished results. Space group, triclinic $P\bar{1}$, $a = 7.3189(2)$, $b = 11.0747(2)$, $c = 11.8449(1)$, $\alpha = 86.61(1)$, $b = 85.24(3)$, $c = 78.50(2)$, $z = 2$, $d = 1.593$ g cm^{-3} (which, interestingly, is identical with that of our (**II**)).

22. D. J. Radanović, S. R. Trifunović, M. C. Cvijović, C. Mariconbdi, and B. E. Douglas, *Inorg. Chim. Acta* 196:161 (1992).

23. TEXRAY-230 is a a modification of the SDP-Plus24 set of X-ray crystallographic programs distributed by Molecular Structure Corporation, 3200 Research Forest Dr., The Woodlands, TX 77386 for use with their automation of the CAD-4 diffractometer.

24. SDP-Plus is the Enraf-Nonius Corporation X-ray diffraction data processing programs distributed by B. A. Frenz & Associates, 1140 East Harvey Road, College Station, TX, 77840.

25. R. B. Roof. "A Theoretical Extension of the Reduced Cell Concept in Crystallography", Report LA-4038, Los Alamos Scientific Laboratory (1969).

26. D. T. Cromer and J. T. Waber. "International Tables for X-Ray Crystallography", The Kynoch Press, Birmingham, England; vol. IV, Tables 2.2.8 and 2.3.1, respectively, for the scattering factor curves and the anomalous dispersion values (1979).

27. H. Okazaki, K. Yomioka, and H. Yoneda, *Inorg. Chim. Acta* 74:169 (1983).

28. R. S. Cahn, C. Ingold, and V. Prelog, *Angew. Chem.* 78:413; and *Angew. Chem. Int. Ed. Engl.* 5:385 (1996).

29. I. Bernal, J. Cetrullo, J. Myrczek, and S. S. Massoud, *J. Coord. Chem.* 30:29 (1993).

30. I. Bernal, J. Cetrullo, J. Myrczek, and S. S. Massoud, *J. Coord. Chem.* 29:287 (1993).

31. D. J. Radanović, M. I. Djuran, T. S. Kostić, and B. E. Douglas, *Inorg. Chim. Acta* 211:149 (1992).

32. D. J. Radanović, M. I. Djuran, T. S. Kostić, C. Mariconbdi and B. E. Douglas, *Inorg. Chim. Acta* 196:161 (1992).

33. D. J. Radanović, S. R. Trifunović, C. Mariconbdi and B. E. Douglas, *Inorg. Chim. Acta* 219:147 (1994).

LIST OF CONTRIBUTORS

F. H. Allen
Cambridge Crystallographic Data Centre
12 Union Road
Cambridge CB2 1EZ
UK
Tel.: +44 1223 336425
Fax: +44 1223 336033
e-mail allen@chemcrys.cam.ac.uk

A. Amann
Univ. Klinik für Anaesthesie und
Allgemeine Intensivmedizin
Anichstr. 35
6020 Innsbruck
Austria
Tel.: +43 512 5044636
Fax: +43 512 5044683
e-mail Anton.Amann@uibk.ac.at

Ivan Bernal
University of Houston
Houston TX 77204-5641
USA
Tel.: 713-743-2718, 2716-2719
Fax: 713-743-2709
e-mail IBernal@UH.Edu

J. C. A. Boeyens
Department of Chemistry
University of the Witwatersrand
P.O. Wits
2050 Johannesburg
Tel: +27 11 716 2342
Fax: +27 11 716 3826 or 339 7967
e-mail jan@hobbes.gh.wits.ac.za

D. Braga
Dept. of Chemistry
University of Bologna
Via F. Selmi 2
40126 Bologna
Italy
Tel.: +39 51 259555
Fax: +39 51 259456
e-mail dbraga@ciam.unibo.it

P. Comba
Anorganisch-Chemisches Institut der
Universität Heidelberg
Im Neuenheimer Feld 270
69120 Heidelberg
Germany
Tel.: +49 6221 548453
Fax: +49 6221 546617
e-mail comba@akcomba.oci.uni-heidelberg.de

T. Cundari
Dept. of Chemistry
Univ. of Memphis
Memphis TN 38152
Tel.: 901 6782629
Fax: 901 6783447
e-mail tcundari@cc.memphis.edu

T. Koritsánszky
Institut für Kristallography
Freie Universität Berlin
Takustr. 6
14195 Berlin
Germany Tel.: +49 30 8383450
Fax: +49 30 8383464
e-mail tibor@chemie.fu-berlin.de

C. Krüger
Max-Planck-Institut für Kohlenforschung
Strukturchemie
Postfach 10 13 53
45466 Mülheim an der Ruhr
Germany
Tel.: +49 208 3062174
Fax: +49 208 3062989
e-mail krueger@mpi-muelheim.mpg.de

J. F. Ogilvie
Centre for Experimental and
Constructive Mathematics
Simon Fraser University
Burnaby BC V5A 1S6
Canada
Fax: 604 2915614
e-mail ogilvie@cecm.sfu.ca

Eiji Ōsawa
Dept. of Knowledge-Based Engineering
Toyohashi University of Technology
Toyohashi, Aichi 441
Japan
Tel.: 81 532 446881
Fax: 81 532 485588
e-mail osawa@cochem.tutkie.tut.ac.jp

B. T. Sutcliffe
Dept. of Chemistry
Univ. of York
York YO1 5DD
UK
Tel.: +44 1904 432515
Fax: +44 1904 432516
e-mail bts1@york.ac.uk

INDEX